COWS AND POULTRY IN INDIA

Their Care and Management

COWS AND POULTRY IN INDIA
Their Care and Management

SECOND EDITION

by
E.B.J.

BIOTECH
2019
BIOTECH BOOKS®

2019 Reserved
ISBN 978-81-7622-097-2

Published by
BIOTECH BOOKS
4762-63/23, Ansari Road, Darya Ganj,
New Delhi-110002 (India)
Phone : 011 4300 3222
E-mail: biotechbooks@yahoo.co.in

PRINTED IN INDIA

Preface to Second Edition

This book has been so kindly received that I am encouraged to bring out a second edition containing some valuable hints which I have recently obtained, and also a few remarks about poultry, which I hope, will be acceptable to my readers.

The first edition has been extensively used by other authors, who have taken extracts from it, but I thrust I may be pardoned for saying that it has retained its position in public estimation, as there is still a large demand for it.

I insert a few copies of reviews and notices, as they may be interesting to friends who take a kindly interest in the work.

Reviews

Cows in India

No one can say that a good practical work on the management and treatment of cows in India has not long been called for. It is a real want, and is conclusively shown to be so by the number of anxious Anglo-Indian housekeepers crying out for information from all parts of the country.

How should cows be fed ? How much milk should one expect ? How much butter from so much milk ? These and a host of others are queries one hears almost every day.

The whole question appeals feelingly to every member of the household. The crying baby, the anxious mother, the fastidious father, wife, the bachelor, are all interested, for they not all turned with loathing from cow and looks of bazaar milk and butter ?

The "main difference" we once heard it said, "between the table of a bachelor and of a lady-governed house, is that on the latter you will frequently find admirable milk and cream and butter, but on the former, never." This dictum must be taken with a grain of salt; but there is a substratum of truth underlying it. The bachelor, if he be a working bee and not a drone, has not (or at any rate he says so) leisure enough to oppose with patient firmness the obstinate roguery of the combined forces of cook and *gwala*. He consequently, he says, prefers indifferent milk and to eschew butter altogether, rather than fight the battle which, as likely as not, may end in discomfiture to himself in spite of the expenditure of time, money, and temper.

There is a point, however, he overlooks, and that is his own want of knowledge on the subject. Ignorance is a poor arm to help him to combat experienced rascals; but the leering *khansamah*, and the apparently half-witted, but really shrewd enough *gwala* are taken aback at once when met with a few meaning statements they cannot in reason contradict.

Ladies have not the excuse of want of leisure but the want of knowledge places them at such a disadvantage that the butter question is an up-hill struggle, and only the staunch ones have the pluck to persevere till the necessary experience is gained. There is no question of the uphill fight. E.B.J. writes feelingly when she says:

"The *khansamah* prefers that milk and butter be bought, as not only does he get a percentage on all bills, but also a fee from the milkman for his patronage. The cook likes writing four ounces of butter in his account, using two, and taking two home to his wife. Careful housekeepers will measure the milk and butter, and make themselves disagreeable in many ways, so all table servants set their faces against keeping cows, and, if they have any hand in the matter, will either endeavour to make the cows a failure, or, if they cannot do this, a very expensive luxury."

E.B.J.'s statement, that "a knowledge of the English dairy is of little assistance in this country," has much truth in it. A knowledge of the treatment of cows in England, and of the working of the dairy, if properly applied, cannot fail to be of value, but to take an obstinate stand upon this knowledge is worse than useless.

We remember that when sent to a riding-school with a number of our fellows, there was a boy among us, who having ridden–as he said–before he could stand, was inclined to regard the riding-school course as a farce, and so much waste of time. When it came to his turn to be mounted, he leaped nimbly to his seat, whence he surveyed the riding-master with a look as much as to say: "Now what do you think of that ?" He looked for a compliment no doubt, but was met by a backhander he has never forgotten to this day, for he reminded us of it only a few weeks ago.

"Ah! my young friend," said the master, "I see I shall have more trouble with you than with all the rest of them put together. It is easy enough to teach a boy to ride, but when he thinks he knows all about it, and has to be untaught before he can begin to learn, it is a very different affair."

So a knowledge of cows in England is a good servant but a bad master. We knew a lady who declared it was all nonsense saying a cow could not be milked without allowing the calf to take a few sucks. She separated the calves from the cows she purchased at the outset; the cows would not be milked, and made themselves ill with fretting; the calves would not learn to drink, and promised to die of starvation, and then the owner gave up cow-keeping in disgust. We intend presenting her with a copy of E.B.J.'s book, and hope it may induce her to make a fresh start.

The *mash* diet recommended by E.B.J. in her chapter on food can hardly be improved upon, and we strongly advocate its adoption. The following remarks, too, are worth recording: "Notice should be taken of the condition of the cow and quality of the milk, and changes in the food made accordingly: if the milk is poor and short of cream, increase the cotton seed: if the cow is weak and thin, increase the gram: if her digestion is not good, increase the bran: if she is a small eater, which many of the best milkers are, lessen the *bhoosa*, as otherwise she will leave much of her *mash* uneaten and lose good food," &c., &c. — *Reviews appeared in the "Asian", March 1888.*

Cows in India: Their Care and Management

By *E.B.J.*

This little book consisting of about 80 pages has evidently been written by one who has a thorough practical aquaintance with her subject, We say "her", because we gather from a passage in the introduction that the author is a lady, although copies of the work are to be procured from Lieutenant-Colonel James, Simla, All scientific terms have been eschewed by the authoress, who writes in a plain, simple, homely style that a child could understand, but the book is one that is written for children of a larger growth, and from its pages almost everything in connection with the keeping of cows may be gleaned. Among the salient features of the work are: Points to be considered in purchasing Cows; Food; Milk and Cream; How to make Butter; Calving; Churning; Bulls; Management of Cows when Calving; Complaints; Indian Breeds of Cows; and many other items germane to the subject in hand. Good milk and butter are comestibles of the highest value everywhere, but especially is this the case in India as we too often know to our annoyance, and our cost. To give an example of the author's style we cannot do better than quote her opening remarks on "Points to be considered in purchasing Cows." "A cow," says E.B.J., "should be loose-limbed &c., &c.

— *Review in "Poona Observer" September 1890.*

Contents

Part II: The Management of Fowls in India

Part I

Care of Cows in India

Care of Cows in India

POINTS TO BE CONSIDERED IN PURCHASING COWS

A cow should be loose-limbed with a large body. Her coat should be soft and silky, and her tail thin and flexible, with a rich and glossy (sometimes wavy) bunch of hair at the end. Her horns should point backwards, not forwards. She should have short legs, wide hips, and be long from the hips to the hocks; her hams should be wide apart, as if nature had intended her to have a large udder when in milk. Even when a calf she should have loose skin about her udder, and her teats should be well-developed. In a heifer the milk-vein is not visible: but when a cow is about to calve, if she is going to be a good milker, it will be very apparent, and will have the appearance of a cord winding round the lower part of the belly. Do not be misled by a fat, sleek cow–she is probably a gross feeder, and what she consumes runs to muscle and fat. The limbs should be loose and large, and the flesh on them inclined to hang downwards. If a cow or heifer is neatly built with nicely rounded limbs and a tendency to fat rumps (not hams), she will make good beef, but will not milk well.

Temper

A good cow should be slow and lazy in her movements, and should look motherly; brisk, frisky cows are seldom good milkers, and are, as often as not, vicious and troublesome at milking-time. A good milker is generally of a dull, quiet temperament, only savage if her calf is frightened. Some are so good-tempered and easy-going, that they will not get excited even if strangers approach their calves.

Colour

Dark brown and red cows are generally healthy strong cattle. There is an idea in England that the milk of a red cow is the sweetest, but I have not found this hold good in India. Red cows have generally good digestion, and are therefore healthy. A very common colour in India is a greyish white; at some seasons a cow of this description will be quite white, and at others a dark grey. It is a colour not confined to any particular breed. I have known some capital milkers of this colour, although, as a general rule, such cows are not good milkers, unless the grey takes the form of pie-bald; when it does, the cow will probably prove a very good one. A creamy white is always a good colour, as it is said to denote richness of blood. The milk is generally very rich, and produces a large quantity of butter. The mouth, the inside of the ears and the horn of the hoof will all be of a bright yellow, and if the hair be also fine, the cow is possibly a large milker as well as a rich one. A pearly white of a bluish hue, with but little yellow in the skin, is to be avoided; the milk will be poor and bluish.

Quality of Milk

Every cow gives milk of a different quality and colour, and these must be taken into consideration when selecting. If the cow is in milk, take a little in a wineglass and look at it; if it is of a bluish hue, the milk is poor. Poor milkers often give a large quantity, but containing too much *serum* to be of use for butter or cream, but such milk is exceedingly good for young children. For dairy purposes, the yellower the milk, the better; and therefore a cow giving yellow milk is considered more valuable than one giving blue milk. Many cows giving rich yellow milk make up in quality for want of quantity. I am no advocate for cows giving fabulous quantities of milk; it is generally very poor, and the cow is probably delicate, as all her strength runs into milk, and sometimes such cows leave off calving early in life. A cow that gives eight seers is a first-rate cow for all practical purposes. I mean eight full seers, or eight quarts by measure; the milk will probably be good and rich. Seven seers is the usual quantity given by good cows, five by ordinary, three by Hill cows, and from two to three by country cattle.

Blind Nipples

A cow with blind nipples should be avoided, for although cows with this blemish are generally of the best milking breed, it shows a tendency to gatherings in the udder, which are very troublesome, and often render a cow quite useless.

Scarred Cows

Do not reject a cow because she is scarred about the face, neck and shoulders, as such scars only show that she has had boils; which have been severely, but effectually, treated with a hot iron.

Age

For many reasons it is better, when buying, to select a young cow, although an older one will have milk of better quality and more of it. The young cow will always sell well after you have had her six or eight years, while the old one may fail to calve again, in which case she is only worth the price of her skin, from four to eight rupees. Again, an old cow is subject to many difficulties which will be detailed under the head of "*Calving*," from which the younger cow is exempt. A cow just going to have her second calf is the best to buy, as she has got over the youthful difficulties of her first calf, and will then give her full supply of milk.

Judging a Cow's Age

A heifer begins to lose her milch teeth when she is three years old, and by the time she is five, her new set is complete. For some time the new teeth remain white and clean, so that from five to ten it is difficult to fix the exact age by the teeth alone. After ten the teeth begin to get long and dusky. Therefore after five years we must judge by other signs. A heifer's hoofs are short and upright, and have very small heels, and the horn of the hoof is transparent and soft; but as she grows older, the hoofs become hard and corrugated, and grow outward or inwards in a saptula shape. In an old cow, the horn at the heels has grown down into a conical shape, the ears have become thick and hard, the mouth broad, and the hairs on it coarse. If you look at an old beast of twenty, you will see all that I have described in an exaggerated form. A cow is at her prime between five and ten.

Judging How Many Calves a Cow Has Had

The value of a cow depends so much on the number of calves she has had, that no reliance can be placed on the statement of the seller on this point, and the buyer must trust to his own judgment.

The best way of judging is by the figure. The hips become more prominent, the belly lower, the skinny parts of the body lengthen and hang, and the udder becomes more raggy after each calf: A cow carrying her first calf shows her condition but little, but an old one decidedly; so, with a little experience, it is not difficult to determine about how many times she has calved.

It is impossible to tell how many calves a cow will have. Some have been known to produce as many as twenty and some do not give more than five. When purchasing, you may calculate on eight, as this is the average number. After her eighth calf, a cow is supposed to be of no market value, although she may have two or three more, in which case the young ones will be strong and healthy, and the quality of the milk excellent, so that an old cow is not to be-despised if you can get her cheap. Not only will her milk be rich, but in all probability she will milk for a long time, as she will not be in such a hurry to calve again as a younger cow would be. But the disadvantage of an old cow is that she is apt to break down at the time of calving; and, being old, she will suffer for days from after-pains, which may interfere with the flow of milk.

If you ask a native how many calves a cow has had, he will generally say four! This is a safe answer, for it is as good as saying that she is neither young nor old. Under four calves, it is easy to judge, as the cow will have a youthful look, and the teeth, &c., all help to tell her age. After eight calves, it again becomes easy, as then the age is apparent; but from four to eight it is difficult to judge with accuracy, but the cow is then in her prime.

To Judge a Good Milker

If the cow you are looking at is in milk, have her milked, and even if she is nearly dry, you can judge by the breadth of the stream whether she is a good cow. Abroad stream shows a good milker, and a thin, poor stream a scanty milker. There is a particular sound which cannot be mistaken when a good milker is well milked; the milk rattles against the tin with a steady rush; anyone who has seen cows milked at home, will recognise the sound; when the supply is scant, the passages are crude and small, and the milk cannot flow freely.

Dry Cows

It is advantagous to buy a cow before she calves, as you then get the whole of her milk; you have also the management of her at the time of calving, and if she is in skilful hands she will give more milk than she would do if mismanaged. It is also well to get her accustomed to her attendants and her stall before she calves, as any change of quarters afterwards will make her fret, and this will at once tell on her milk. If the 'calf is not very young, any check of this kind will lessen the milk permanently. So it often happens that the purchaser of a cow in milk complains that the cow does not give as much as he was led to suppose she did, the reason being that the milk has been lessened by the fretting of the cow on changing hands.

Dry cows are much cheaper than those in milk; the cost of keeping them before they calve will not exceed the difference in price between a dry cow and one in full milk, and you will be fully repaid by the improved quality of the milk, due to care and proper feeding.

Many persons do not care to be burdened with a dry cow, and sell it for anything they can get, for getting that it will repay them to have a few dry cows, who will in due course come into milk in turn with those already milking, and that the rest and care will make them all the better milkers when they again commence. This impatience on the part of his master suits the native servant exactly. He takes care that the dry cow shall be sold for a mere song, and that a high price shall be given for the cow to replace her. He and his friends will benefit by both transactions.

Length of Time a Cow Milks

As a rule a cow is in her *best* milk for six months (some cows for eight or ten months). This is the accepted time, so when you buy a cow, judge how old her calf is, and how much of her *best* milk you have lost: it ought to make a great difference in the price.

Value of Calves

When purchasing a cow in milk, be particular that the calf is as well-bred as its mother, and note its sex, for a bull-calf is really more valuable than a cow-calf, for after three years it will probably be worth from sixty to eighty rupees, or even more, while a cow-calf, after the same time, will seldom fetch more than forty or fifty rupees. Some people look upon calves as valueless, and give them away to their servants–a most foolish practice, as it gives the servants an idea that all profits accruing from their masters' cows are the servants' right, and, accordingly, they make it their business for ever afterwards to stop the sale of calves.

The following are the points of a good bull-calf:

1. He should have loose skin running from his chin down his dewlap, and round his chest and along his belly; his eyes should be far apart, and his forehead broad; his nose short and turned up, his joints large and looking swollen. The size of the joints is an indication of his eventual growth, and the better bred he is, the more ungainly will his joints be. The shorter his neck the greater will be his strength. His hair should be soft and long, and rather curly. If he is in good health, his nose will be wet and shining. An ill-bred calf will have a narrow head, long ears, small eyes–rather close together,–long hams, small joints, and will taper away at the feet like a goat.
2. A cow-calf should have much the same points as a bull-calf, but her head will not be so broad, nor her neck as short; and, instead of the long dewlap, she will have, if of a good milking breed, plenty of yellow wrinkled skin between her legs, where the udder will be in time to come; and she will have well-developed teats, even when only a few hours old, and the best milkers always have silky hair.

KEEPING COWS

Having purchased your cows, it is necessary to turn your thoughts to their keep. This is a thing but little understood; it is generally left to the supposed knowledge of the *Gwalla*, who has his own ideas, founded on the traditions of his fathers for many generations back. As natives believe that a son must be inferior to his father, it follows that the former must not try to improve upon the work of the latter; what was good enough for the one must be good enough for the other; consequently your *Gwalla* will resist interference, and if you introduce English methods of dealing with cows, he will have the satisfaction of seeing you fail, for, as I have said before, cows in this country require very different treatment to what is considered orthodox at home. Native superstition must also be taken into consideration, for the cow is a holy animal, and many are the religious observances connected with her. Her feelings must be respected by a good Hindoo, who will not insult her by looking at her teeth, even as it is not proper for a young woman to show her teeth by smiling before men. Again, if a Hindoo buys a cow, a *pooja* or religious ceremony has to be performed at the peg, which must be a new one, to which the cow is to be tied; this makes it holy, and ensures a blessing on the owner as long as the cow remains tied to that peg. So great is their reverence for cows, that I have always found Hindoos most faithful servants to me, because I have reputation among them for being fond of cows, and treating them well. Once, when driving, I narrowly escaped being badly hurt–the cart was smashed, but I was uninjured. The Hindoos attributed my escape to my love for cows. It is strange that in a country where cows are worshipped, custom should have made the keeping of them a monopoly in the hands of a low and bad caste, called *Gwalla*s, who have, from time immemorial,been thieves and rogues. One would have thought that the holy animal would have been tended by men of a holy caste, such as the Brahmins. The whole business of taking care of the cows, and milking them, and making butter, *ghee*, &c., is in the hands of the *Gwalla*, and consequently he looks upon the animals and their produce as being, to a great extent, his own property, and all profit as his exclusive right: and he will not, if he can help it, allow his master to benefit in any way by the dairy beyond obtaining what is required for ordinary use. He is in league with all the other *Gwalla*s in the place, and will, if he dare, dry your cows, steal your milk, starve your calves, and, reduce you to the necessity of buying milk in the bazaar. He has no conscience in the matter, for he says that God has made him a *Gwalla*, and that, therefore, he has divine authority for appropriating the produce of the holy animal, to which you, as an earthly proprietor, have but a very inferior claim. I therefore advise my readers never to keep a *Gwalla*. With natives custom is everything; it is with their religion, so you see in a Hindoo country starving cows and ill-used calves,. because custom has made the wretched creatures over to this worthless caste of *Gwalla*s, who are cruel and avaricious, and whose only object is to make money out of the animals. Not only are the cows neglected and ill treated, but the calves are starved until it is really pitiful to see such abject skeletons; sheer want forces them to eat grass long before they are old enough to digest it, and the consequence is shrivelled limbs and distended stomachs; the poor beasts grow up with difficulty, and only get a little stronger as they come to maturity, when they commence producing another generation of starving calves, for as the *Gwalla*s sell every drop of milk they can get, the calf is only allowed to suck what the man cannot extract. The country cow would be a much finer animal than it is if the calf were properly fed when young. It is much more hardy than any other animal of its kind, or it would never have survived the treatment it has received from time immemorial. All well-bred draught and carriage bullocks are allowed the whole of their mothers' milk; hence the beautiful animals which they generally are. They do not belong to *Gwalla*s.

From the above it will be understood that the first step towards success in cow keeping, is to get a servant who is not a *Gwalla* to tend them. If you keep one or two cows only, perhaps one of your servants may know how to milk, in which case he will be glad to look after them for a little extra pay. A boy on a couple of rupees a month will graze them. But if you have many cows, you must have a servant for their exclusive charge, and there are plenty to be had who are not *Gwalla*s. The *Aheers*, who are country folk and in the habit of keeping cows and buffaloes, are the best of the *Gwalla* caste, and sometimes make good servants. Mohamedans also do wel1, as they have no prejudices where cows are concerned, and are consequently ready to carry out the wishes of their masters. They have no divine right to your milk and butter as the *Gwalla* has, but neither have they any reverence for the cow, and would gladly turn her into beef.

When you have arranged for a good cow-keeper, have a large *nâd*, or earthen basin, built up into a feeding trough; it should be raised from the ground about one foot to enable the cow to eat with ease; it should be strongly built and thickly plastered. or the cow will knock it about with her feet and horns, and it should be placed in a nice bright place, where she can get the morning sun. A similar *nâd* should be placed inside the cowshed, where she will have to be fed on wet and windy days; it is also useful at night for the *bhoosa*, which is trampled on and spoiled if left on the floor. An earthen feeding trough is preferable to a wooden one, which falls to pieces if allowed to get dry, and swells if kept wet; when the *nâd* gets unpleasant it can be easily changed. It should be well washed out after every meal, and then left to dry before the next meal is mixed in it, for cows are dainty feeders, and will not eat their food if it is in the least sour, as it will be if the *nâd* is not kept perfectly clean; if by any chance they do eat sour food, it is very apt to make them ill. Poultry should not be allowed near cows, as the fowls scratch in the cows' food, and their droppings spoil it, and, if eaten, cause severe gripes. In the cold weather cows should have clothing, as a milk-cow feels the cold very much, and her milk is affected by it; they will not keep in good case if cold. In the hot weather they should be tied out at night, as a close stall will lessen the milk. In the rains an open shed is best, as it gives shelter from the rain, as well as allowing a free current of air. At this season of the year it does not matter how wet your cows get during the day: in fact, it is well to send them out in damp and rain, which cool them and help to cure their prickly heat, from which they often suffer so much as to appear quite mangy. Grazing in the wet often cures this altogether. The long juicy grass in the rains also increases the quantity of the milk. You will get no more butter in the rains than you did in the dry weather, but the colour of the butter will be yellow; green food produces yellow butter and a finer flavor. A good supply of drinking water should always be at hand near the cow-yard, so that the animal should have a good drink at stated times during the day. Cows are large drinkers. The quality of the water should be carefully looked too. A milking cow is so thirsty during the dry weather, that she will drink greedily out of any dirty *jheel* or puddle she comes to. She prefers clean water but if she cannot get it, she will take dirty water, as drink she must. If she has a plentiful supply of clean water at home, she will prefer it to dirty water abroad.

Drink

In these days a great deal is said about village milk from badly-fed cows producing typhoid fever: it is also asserted that dirty water is often mixed with bazaar milk; but if more attention were paid to the *water* cows *drink*, I think the doctors would be nearer the mark.

We all know to what filthy uses natives put *jheels*, and *jheel* water must be unfit for milking cows to drink. No native would mix dirty water with his milk. as that would change the colour and spoil the sale of it.

French Dairy farmers say if a cow be given warm water to drink instead of cold, she will give one-third more milk, This is certainly the case in the Hills, and should always be done there.

Forcing Supply of Milk

Some people advise giving cows large drinks of warm water, with bran or flour sifted into it to make it tasty, so that they may drink eagerly, and give a larger supply of milk. This is sometimes done at home. and is a practice resorted to in this country when a cow is for sale. Large quantities of salt are also given to produce thirst, so that she may drink to excess. I see no reason for this anxiety to increase the quantity of the milk unless it is for sale. as I have found by experience that milk increased by drinking is not as rich as that increased by food, and that an unnatural increase in the amount of milk is always followed by a reaction and a proportionate lessening; for instance, if your cow is giving four seers, if you force the supply up to four and a half seers, after a few days it will fall to three and a half: of course it is different if a cow is falling off in her milk; then giving her cheretta sticks chopped up in her food, or a handful of tea leaves in her mash, will often restore it again for a time. As a rule, all changes in food and water will affect the milk, and when a cow is doing well, it is best not to make alterations. All journeys, moves and changes lessen the milk.

Some cows suddenly stop milking for no reason whatever, except that they have been milking well for many months; giving them ¼lb of tea boiled in a bucket of water, with *goor* or *suttoo* mixed in to make them drink it, will sometimes bring back the flow of milk. Should the milk suddenly stop from any other cause, the udder will swell and become inflamed, and tender to the touch.

Lucerne Grass

Lucerne grass, or clover, is bad for cows in India, as it heats the blood and dries up the milk and makes the animals want to calve long before the ordinary time; you may give it to a young heifer if you want her to calve early, but as it tries a cow's constitution if she calves too young, and the calf is likely to be weakly, it is better not to hurry nature. If left to herself, a heifer will not get in calf for sometime after she is old enough to do so, as she is too timid to allow any strange cattle to come near her.

Mitha

Natives have an idea that it is a very beneficial fodder for cattle, and grow it in large crops for that purpose. It is too heating for milking-cows, and gives the milk and butter a bad flavour.

Green Food

Carrots are excellent as food; chop them up in the mash; they sweeten the flavour of the milk, and add to the quantity of butter. Sugarcane and beet-root are also wholesome and sweet in the spring (February). In most parts of India you can buy young wheat or barley (called *kood* in Bengal and N.W.P., and *kuhseel* in the Punjab); it is a splendid food for cows, and makes the milk as rich as English milk, and the butter a bright yellow. On the days: you feed on green barley, the milk will produce one third more cream than usual; the barley also acts as an alternative, and helps in changing the cows' coats, which they. shed at this time. A bunch of cabbage-leaves occasionally will do no harm, but avoid cabbages in large quantities, and also turnips and onions, as they give the milk a rank flavour. Silo grass is good food for milking cows, and they do not mind the smell of it. The "Riana Luxuria "grass also produces beautiful milk.

Winter Coats, Lice and Ticks

While cows are changing their winter coats, they sometimes look mangy, and the new hair is so irritable that they will rub themselves sore. During the winter, when the coats are long, they are much troubled by lice of a bluish colour and very small. It these are not removed, the cow becomes quite sickly from the great irritation. The cure is simple–phenyle and water rubbed over the animal. The insects are so small that they easily escape detection unless the coat is carefully examined–when you observe a cow, rubbing herself constantly, look for parasites. Ill-fed calves often get them in large numbers, and become so weakened that they even cease to exert themselves to scratch. Dog-ticks also attack cows, but are easily picked off. They cling about near the udder, and under the elbows of the poor creatures, in places where they can neither scratch nor rub them off, and a cow will put down her head and sigh with relief when her attendant removes them.

FOOD

A cow in milk should daily have a mash of the following ingredients, divided into two feeds, one given in the morning, and one in the evening before milking time:

Bran (*bhoosee*)	1 seer
Gram (*chunna*)	1 seer
Cotton Seed (*bunaula*)	1 seer
Oil-cake (*khullee*)	1 seer
Chaff (*bhoosa*)	2 seers
Green Grass, chopped	4 seers

The above is sufficient for most cows, but in the case of a very large cow or a large milker, the feeds may be increased. Notice should be taken of the condition of the cow and quality of the milk, and changes in the food made accordingly; if the milk is poor and short of cream, increase the cotton seed; if the cow is weak and thin, increase the gram; if her digestion is not good, increase the bran; if she is a small eater, which many of the best milkers are, lessen the *bhoosa*, as otherw1se she will leave much of her mash uneaten, and lose good food. Green grass gives colour and richness to the milk; cotton-seed produces rich, but not yellow. butter; oil-cake helps to produce milk, and by natives is the only food given besides *bhoosa* and grass; it is wholesome and sustaining, and, when, soaked in water, makes the dry *bhoosa* palatable. *Khullee* is the oil-cake which is the refuse of the seed when the oil has been pressed out; the gluten and germ of the seed are left in it, but the oil is gone, and it has been so thoroughly crushed that it is easily digested, and, when soaked, becomes like porridge. There are many kinds of *khullee,* as oil is made of so many Indian seeds. The proper kind for cows is that made from the *till* or *ulsee* seed (*Desamum Orienlale*), a plant bearing an oily seed. Natives are fond of this oil-cake, as it is sweet in flavour; it has an oily smell and is of a chocolate colour, and, if old, is as hard and dry as a brickbat: before it is broken up it is generally in the shape of a large cylinder, this being the form of the press. No other kind of oil-cake should be given to cows. Other kinds are heating, and make cows ill, and give the milk a nasty smell. As native servants often try to impose on their employers by buying the wrong kind of oil-cake, owners of cows should make themselves acquainted with the look and smell of that made from *ulsee*. It is very easy to distinguish, as the inferior kinds are cheaper, blacker, harder, and have an offensive odour, as they are made of *kussum* seed, mustard seed, poppy seed, *mahow* seed, and a dozen other kinds; but they are all pungent, and disagree with cows. On account of their cheapness they are used to feed bullocks. About four hours before it is required, the oil-cake should be broken up and soaked in as much water as is necessary for the mash. The gram should also be soaked with the cotton-seed, which is by nature very hard. When the gram is fully swollen and the oil-cake dissolved, the bran and *bhoosa* should be mixed with the water in which the gram has soaked, the cotton seed and the *bhoosa* should be added to the mixture together with green grass and a handful of salt, and the mash is ready. Care should be taken not to soak the oil-cake for more than four hours, or it will turn sour. The green grass should be picked or cut from some cool spot near a well or running water, or under a bank; it should be washed and chopped and mixed into the mash. Green grass is very necessary in the hot, dry weather, when the cows cannot find green food; the cow-keeper can always manage to get sufficient from the garden or other well-watered place.

In places where good oil-cake is not to be had, it is as well to substitute *atta*; instead of soaking it in the way recommended for *khullie*, pour boiling water over it, stir it into a porridge and mix it in with the other ingredients, only being careful it is not allowed to go sour or the cows will not touch their food.

In the Hills, it is always difficult to feed milk cows, as in some hill stations *bhoosa* is not to be had–in others grass; so all kinds of plans have to be adopted to keep up the cows' mash. In the Hills natives are fond of soaking all food in a bucket; and allowing the cow to drink it off at a draught, but this cannot be good without the addition of chopped grass or *bhoosa*.

> Feed at 7 A.M. in the cold weather, and at 5 A.M. in the hot weather, and always before milking, as the rush of milk on an empty stomach tries the strength. After the cow has been fed and milked, her calf should be tied up, and the cow taken to graze. It is well to send them out at all seasons, as the exercise is beneficial, and the dry roots they pick up in the hot weather are nourishing, although dry–*doob* grass is full of nourishment, even when not green. I have often heard it stated that it is dangerous to allow cows out to graze, as they are apt to eat all

kinds of garbage, which affects the purity of their milk. This is a popular fallacy, as a cow is by nature a very clean feeder, and will not eat improper food unless starved or educated into such vice. I know from my own experience that well-fed cows will refuse their food if it has not been cleanly prepared and mixed, and the *nâds* daily washed out with hot water; I have seen them standing quietly whisking their tails, and not touching their mash, simply because it smelt sour, or had been mixed with dirty water, or the green grass had been gathered from the low land near the river, and smelt fishy and rank. But, as exceptions prove a rule, I will mention the case of a lady who had a pet cow which had been taught to eat whatever her mistress offered her; I have seen her eat a plateful of curry and rice! But this is accounted for by her having been petted from her birth, and fed by hand on everything–cakes, bones, biscuits, fish–until her taste was quite vitiated; I believe she would have eaten and enjoyed a beefsteak if it had been offered to her!

If a cow does not get salt, she will often pick up dry bones and crunch them–for the same reason she will lick a white-washed wall; but I am sure that a properly-fed cow that gets salt regularly, will not touch dry bones. Many people advise giving a large quantity of salt to produce thirst, which makes the cow drink largely and increases the milk, but, as I have said before, the milk will not improve in quality, and the supply in the end will be decreased, and perhaps the cow injured. In some parts of India cows are fed on dreadful filth–horses dung and stable litter-particularly that from Artillery and Cavalry stables, and other well fed studs, where the animals receive much grain. The natives say that it gives the milk a rich creamy look, and in no way alters the flavour; but it is only done in parts of India where food is difficult to get, and expensive. The poor cows are made to eat it by cruelty and starvation, and it has to be prepared and made palatable before they will touch it–no cow will do so until want become her master. But I have no object in exposing all the blackguard tricks of *Gwallas*; we all know that bazaar milk may be anything! Even in a Christian land like England, we must not enquire too closely into London milk, or we may perchance discover that it is not milk at all! It is well-known by people who understand cows that they are very clean feeders. I may say more; they are fastidious feeders. If you put a bit of bread to your lips and offer it to a cow, she will not touch it. Few cows will eat bread at all until they have, become accustomed to the flavour, as they do not like the smell of the yeast. I think the above will show that your well-fed and well-cared-for cow can go out to graze and exercise without the least fear of her picking up anything harmful to herself or to her milk.

When your cows return from grazing, they should have their evening meal, which should be prepared in the same way as the one in the morning. After their dinner they should be milked. It is well to allow the calves to be loose while the cows are out, as too much tying up cramps their limbs and stops their growth; they have to be tied up while their mothers are at home, or they drink all the milk.

MILK AND CREAM

As soon as the milk of a newly-calved cow has lost its bright yellow colour, it is fit for use for tea, coffee, and puddings. As long as it is yellow it may turn if boiled for coffee or cooking, and it does not look nice for tea, and also has a sweet taste, but for cream it is excellent, as it is as rich as it can be, and butter made from this new yellow milk will be of good colour and delicious flavour. Natives covet this new milk, as a special dish is made of it which cannot be made of other milk which is not so rich. English people as a rule give it to their servants, or let the calf drink it. I let my servants have it for the first two days and then I use it for butter until it has acquired its proper colour. The calf should not have too much of it, as its richness acts as a purgative, due to the presence of colostrum, and, for the same reason, it is bad for children. The milk of a young cow is poorer than that of an older animal. With a first calf it is generally very poor, and produces but little cream and butter; but it is particularly good for babies and young children, as it is easily digested. While the calf is very young the milk will be poor, and will get richer as the calf grows older, and will have more body and better colour. As a cow goes out of milk, the butter obtained will not lessen as rapidly as the milk itself does; the smaller quantity of milk will, on account of its richness, give a larger proportion of butter, and sometimes, even when nearly dry, a respectable part of butter will be yielded.

The milk of every cow differs in body, quality, and colour. There is the yellowish, creamy milk, which contains a large proportion of the fatty substance necessary for butter. Then there is the thick, heavy, white milk, which contains a great deal of casein, suitable for cheese, junkets, curds, &c., and the thin, bluish milk, which is sweet, and nice, and generally very plentiful, but does not produce much cream or butter. This is the most common kind of milk produced by native cows.

Watered Milk

Milk which has been watered always takes a bluish look, which can be easily detected by looking at a little in a wine-glass, as it appears more transparent than pure milk. But the best way of testing the milk is to taste it; a little practice will enable you to detect watering. Water has a wholly different flavour and feel to the tongue; it is harsh and tasteless, while pure milk is soft, sweet and smooth. The glass bulb or lactometer for testing milk is not always to be relied on; it will show the bluish or pearly milk, which I have described above, to be half water, whereas in reality it is quite pure. The milk of an old cow will read purer than that of a younger one. The difference in the food of cows will also affect the specific gravity of their milk. The long, juicy hill grass makes the milk look like water, yet the cream will rise quite thickly, and the butter come well.

Milk which has been watered will turn sour sooner than pure milk, so if your milk turns sour more quickly than it ought to do, you may suspect it; also, if sour milk be warmed, it will, if it has been watered, curdle up into small feathery curds, while pure milk, when sour, will cake into large solid curds. Another test is to stand the milk for a short time. if the cream comes in a thick crust, it is pure; if, on the contrary, it breaks up on skimming, and will not readily come off the milk, it has been watered, and the skimmed milk will then be of a blue colour, and look like white agate or opal. The process of skimming watered milk is difficult, but it is easy to skim pure milk, as the cream comes thick and solid. Bought milk should always be boiled for safety, as many kinds of germs which produce disease, grow well in milk. Boiling for ten minutes kills these germs. But it must be remembered that even boiled milk is a good medium for the growth of pathogenic germs. It should therefore not be left exposed to dust, or offensive smells, or in dirty vessels.

Keeping Milk Good

Pure milk keeps good much longer than is generally supposed. A few points should be observed. Nothing turns milk so soon as china jugs or crockery, which retain heat and are close. All vessels to contain milk should be copper well tinned, and should be rubbed with dry wood-ashes until every particle of grease is removed, then scalded with hot water and put in the sun to drain and dry; this method of cleaning is easy and effective, and readily adopted by natives who have been shown what is wanted. For setting cream you should have well-tinned copper pans of the following dimensions:

Length	12 inches
Width	7 inches
Depth	1½ inches

The should have shelving sides like the sketch below:

This shape is broad and open at the top, and allows of a dinner spoon passing round to collect the cream, and a number of such vessels stand compactly on a shelf, where they can be placed side by side in a row. When they have been in use for some time, and constant rubbing with wood-ashes has worn off the tinning, they should be re-tinned, and care should be taken that the tinman (*kulaigur*) fastens up the corners, which the heat of re-tinning may have unsoldered. Before use they should be tested by being filled with water, otherwise you may lose some milk by leakage. You should have a double supply of these vessels, so that one set may be in use while the others is being scrubbed, and dried in the sun. It is necessary to arrange shelves according to the amount of milk you have to set. They should be placed in a quiet room which is not a passage; and which can be shut up and kept cool; people passing through a room disturb the atmosphere, which should be perfectly still for the collection of cream. When the weather gets warm, it will be necessary to open out the room at night to admit cool air, but the longer you can keep it shut up the better, as even the breeze from an open window playing on the surface of the milk causes a slight motion, which retards the collection of the cream. No cream vessel should be moved or touched from the time the fresh warm milk has been poured in to the moment of skimming. Some people place their cream vessels in wet sand, which is a good and practical plan, especially in a temperate climate like that of Bengal, where the weather is always a little too hot for cream and butter-making. Where wet sand is used, *pucka* or stone stands with rims or ledges round them should be made to contain the sand; the vessels are placed on them.

Atmospheric changes affect milk very much. Sometimes on a cloudy day the cream will be quite thin, and even if the thermometer is not high, a fall in the barometer will often be followed by the milk turning sour; when a westerly wind is blowing, both milk and cream will diminish, and frost also retards the collection of cream, although a bright clear day without any wind is usually most favorable, but wind can always be shut out by closing doors and windows. The room where cream is set should have wire netting across the windows, or, if this is not procurable, bars should be placed across them to keep out cats and crows. Until the thermometer rises to 70°, butter may be made of unboiled milk, which gives it a finer flavour than milk that has been boiled. If you intend to devote all your milk to butter making, set it for thirty-six hours; the cream will then set on the top in a thick paste, leaving the milk below quite thin and poor, unfit for household use, but good enough for the calves. Cream which has collected for thirty-six hours is such a cakey crust that it is easily removed and makes the best and largest amount of butter. If you wish to use the milk for both butter and household purposes, it is as well to skim off a tin or two, as required, from those more recently set. By boiling down skimmed milk it regains a good body and is useful for many household purposes. Skimmed milk which has stood for twelve hours is good for cooking and may be used in other ways, as well as having produced a good deal of cream for butter. The best and easiest plan is to set the morning's milk, say on Monday; then set the evening's milk next to it; on Tuesday morning skim Monday morning's milk for the calves, and Monday evening's for the household, filling up the empty places with Tuesday's milking, and so on. In this way you as nearly as possible "have your pudding and eat it too."

Skim the cream into the mugs in which it is collected, every morning until you have a sufficient quantity, say the cream of four days, the butter will come all the sweeter and better for the cream and being a little sour, and will keep good longer. By keeping, some chemical change is produced in the cream, which makes the fatty part come away more freely when churned; the butter made from too fresh cream will often turn sour in a few hours, for the acid portion has not been thoroughly separated from the greasy part.

In the Hills, where the climate is cold, and especially so in the winter months, the milk need not be skimmed for two, three or four days, when you get a lovely thick crusty cream, and if you use the copper pans I recommend, they are easily skimmed by breaking the cream at one corner, and tilting up the pan on the opposite corner; the milk will pour out from underneath, leaving the crust quite clear a spoon or knife held across the corner where the milk is running out, keeps the cream crust from falling. In the Hills, I always wash out the pans with warm water, and put the creamy washings into the churn. and by this plan save *all* the cream; it makes perceptible difference in the quantity of butter obtained. In this climate you can churn once a week, or every 10 days, in true English style, but to do this you require a large churn, and the butter will come slowly, requiring hot water poured over the churn, or standing it near a fire, or in the sun. If the weather is *very cold* and the butter will not come, pour a little hot water in, and go on adding till the butter comes. Once commenced; churning should never be stopped till the butter comes, or it makes endless labour.

Butter-making in the hill climate is easy work, and every lady should succeed in making nice butter with a little care and patience.

TO MAKE BUTTER

To make butter, put your collected cream into an atmospheric churn, if you have one. If not; use a home-made one (see "Home-made Churns"). I always use the latter, and have nothing to complain of. Add a few mugs of *cold* water, as it keeps the cream cool during the churning, and also, being a harder fluid than the milk, helps to beat up the butter. English dairymen maintain that there is so much in this beating together of differently weighted fluids, that more butter will be obtained from the milk of two cows, if beaten together, than from the milk of the same cows, if churned separately. Commence churning slowly; the more slowly the butter comes, the better it will be both in quality and quantity. Be careful that the churn is securely closed, and increase the pace gradually. Take care that the natives do not hold the churn with bare hands and legs, which will make it warm, and cause the butter to adhere and become oily; they can hold it quite steadily with a duster or *jharun*. As soon as you feel the fluid becoming heavy and cloggy, churn more slowly, and continue to do so until you find the churn running, easily again, and distinguish the sound of the fluid rattling about inside; then look in to see if the grains of butter are coming, and continue churning *gently* until the butter forms a solid lump at the bottom. While churning, occasionally draw the, handle up to its full length, so that all the particles of butter may be collected, and none left adhering to the top of the churn. If the butter is turned out too soon, that is, before it has become a solid lump, a great deal is lost, as it is impossible to collect all the feathery particles when it has been once taken out of the churn: natives always take it out too soon.

The process of butter-making is very simple; it is churned until it becomes a firm foam called "whipped cream"; then it bursts into little particles leaving the fluid free; then the little particles are knocked together and adhere to one another until they form a solid lump of butter, leaving the fluid clear of all greasy matter. This fluid, or "Butter-Milk," is very rich, and if much water has not been used in the making of the butter, it makes capital Scones, Soda-Chapaties, Suet Puddings, and other things of the kind. They rise beautifully with it if a little bi-carbonate of soda be added.

Your butter being ready, pour off the butter-milk, and give it to the cows, mixed up in their mash; it is a capital thing for them. Wash your butter with cold water which has been standing all night in the *chatties* or *ghurras*; fresh water from the well is generally warm, and may melt the butter. After washing it *thoroughly*, salt it and make it up into pats; butter made in this way will keep sweet and nice for many days; it should be re-washed and salted every morning until the next churning day. If the cream is too fresh before churning, the butter will turn sour very quickly; it should be collected and made. into shapes with a wooden spoon, or with a knife made of bamboo, and should never be touched by the hand or left in a warm place, even for a few minutes.

After the thermometer reaches 70°, butter-making becomes troublesome, as the heat makes the butter oily and soft. Lumps of ice dropped into the churn while churning help to harden the butter, and the churn should be placed in a tub of cold water. You must now make your butter every second day, and skim your cream vessels oftener than you did during the cold weather.

Cream Butter

Cream butter can be made from the cream of boiled milk when the thermometer rises to 75° in the mornings, and 80° during the day, but it requires care and management. The milk will remain good if it is boiled both before setting and again after skimming, and is kept in the proper vessels, but these vessels *must be kept clean.* Want of perfect cleanliness is the great obstacle to success in butter-making, and especially so in hot weather; and without the eye of the "house-keeper perfect cleanliness is not to be obtained in this country, for no native knows a clean tin from a dirty one! The slightest suspicion of sour milk from the day before, even a little greasiness, will turn the whole of your milk into *tire* (an Anglo-Indian term for sour milk), when the weather is hot. It is even necessary to take care that the same spoon is not used for the skimming of last night's and yesterday morning's milk, as last night's milk, when re-boiled, can be used all day, while yesterday morning's milk, though quite sweet, it is only fit for the calves, as it is more than twenty-four hours old, and consequently will not stand another boiling. As the weather gets hotter, the milk should be boiled for a longer time, and oftener skimmed, when milk has been thoroughly boiled the cream rises so quickly, that in very hot weather it need stand for a few hours only, and in this way you will get plenty of very nice butter, and be able to use the milk as well. Plenty of ice must be used in churning, and when the weather is very hot, *i.e.* in May and June, skim the morning's milk at 12 o'clock, boil again and set a second time; the milk will then be fit for use all day, as it becomes richer the more it is boiled, and cannot go sour *if all the pans are clean.*

Should your milk turn sour any day, do not be discouraged, as it has probably done so on account of some dirtiness in the pans or milking-can; or perhaps there is a storm brewing, and the electricity in the air has turned the milk, as it will do in a few minutes. Whenever things go wrong, look for the cause, and, if possible, remove it.

As the weather gets warmer, the milk must be each day more thoroughly boiled before setting, to insure its keeping good. It is wonderful how much can be done with milk in the hot weather by careful observance of the above rules.

When the weather is very warm, the butter will no longer form a solid mass at the bottom of the churn, but will cling all about, and just after churning will look quite oily; plenty of fresh cold water should then be poured in, and as much of the butter-milk as possible poured off, and a lump or two of ice should be dropped into the churn; this will soon make the butter hard enough to take out and make up with salt into a pat. The butter should be kept in an earthern (*mutti*) vessel full of cold water, which should be kept in a cool place; the water should be frequently changed to keep the butter sweet. Another good plan is to put it into a corn-flour box, and the latter into the ice-box. As the weather gets warmer, more boiling has to be done, and more ice to be used. In this way I have made butter all through the hot weather, and can answer for its success.

Before churning at any time of the year, it is necessary to soak all the wooden parts of the chum in cold water for some hours, so that the pores of the wood may be saturated with water. This prevents the butter adhering to the wood.

All wooden articles used for churning or making 'butter' should be of white wood–deal or bamboo,–toon and an all dark, heavy, close-grained woods should be avoided, as the butter adheres to them; and their weight presses it too hard. All wooden moulds should be well rubbed with salt, as well as soaked, in order that the butter may come out of them clear and distinct. When the weather is very cold, coarse cloth may be used to press out. the water from the butter, but it is a messy plan, and not really necessary–a little time and patience and a wooden knife will ensure a good pat.

Butter Made in a Basin

In the Hills, and, in very cold weather, in the plains, a good and simple method of making butter is by beating up the cream in an enamelled iron basin with a dinner spoon, or, better still, a piece of bamboo cut into a whipper–like this:

It soon whips up the cream. For this kind of butter-making, be careful to skim the cream with as little milk as possible, as the thicker it is, the sooner it will whip up; and it should be done in a cool place. Beat very slowly, being careful that all is beaten alike; collection all that is hanging about the basin, so that none may be left still frothy, as that will be butter lost; work it round and round until the butter collects into curdled lumps. If it refuses to turn into butter on account of the cold, you should, as soon as it is beaten into a stiff state, move it into a warm room or into the sun, and go on beating; it will turn into butter in a few moments. This is a capital and simple way of making butter when you have no churn, and have only a little cream, but it would be a very laborious method for a large quantity.

Butter made in this way soon turns sour, as it has not had enough friction to drive all the gas out of it, so, unless the weather is very cold, the butter will turn, and if the cream is not quite sweet, it will also, for the same reason, make sour butter.

Butter Made in a Bottle

Another simple old Indian way is by putting the cream into a fruit bottle with a piece of wood inside. The bottle is corked and shaken violently until the butter forms a lump inside. This, too, can only be done with very thick cream, and in cold weather. Care has to be taken that the heat of the hands does not make the butter oily. It is a capital method of making butter on a march. After the cream has been set, the *khansama* skims it and puts it into a bottle, and shakes it up into butter as he goes along, sitting on the top of the bullock cart; and when the butter has come, he plunges the bottle, butter and all, into a bucket of water, where it remains until it is wanted, and where it keeps cool and free from dust, &c. It is rather a long process, as the cream does not beat up quickly, as it is too confined, but it suits natives, who will sit happily for hours, chatting and shaking and passing the bottle round the circle.

Native Method of Making Butter

The way in which natives make butter from both cow's and buffalo's milk is to reduce the milk to one quarter of its original quantity by boiling it over a slow fire made of *kundees* or *gointa*,– cakes made of cow-dung dried in the sun. They smoulder, and by burning so slowly retain a great deal of heat, and the milk gradually steams away. Natives think that by reducing the milk in this way more butter is obtained. When it is sufficiently reduced (in some cases it is kept on the fire for twelve hours or more) the milk is set, and congealed by a little sour milk or *dahi* being stirred in while it is still warm; early in the morning it is churned; the butter-milk is sold. Though by this method more butter is obtained, the *kundee* fire gives the milk a smell arid flavour very disagreeable to Europeans, though highly appreciated by natives, and which do not quite leave the butter even after it has been rendered down into *ghee*. Another reason why country milk and butter have such a sickening flavour is to be found in the fact that the milking vessels and churning *mutkas* are, after being washed, turned over the smoking *kundee* fire, to preserve and clean them for the next milking and churning.

Butter-making as Carried Out by *Gwallas* and *Khansamahs*

The old Indian way of making butter, and the plan still adopted by most natives, both by professional butter men and by domestic servants *gwallas*, &c., is as follows:

The fresh milk is boiled, then cooled by placing the vessel in which it has been boiled in cold water, and gently stirring it until it is cold; a little *tire* is put in and stirred round; then a cloth is tied over the top of the vessel, which is put away to set. When the whole mass of milk has become *tire* or *dahi*, the churning commences--in the cold weather in a hot place like the kitchen or pantry, or in the sun; in the warm weather out of doors in the dew. The churning should, during the hot season, be done very early in the morning; and, when the weather is very hot indeed, before daylight. In warm weather the milk soon congeals and becomes a solid mass of *tire*; in the cold it requires some management to get it set in time; when it refuses to set, a little more *tire* has to be added, and the jar put in a warm place. Natives never churn until the milk is quite congealed, and smells very sour. The coming of the butter depends entirely on the temperature of the milk, and it requires an experienced butter-maker to tell whether cold or hot water should be added. The practised eye of a constant butter-maker will decide this at once, as the following anecdote will show. I once tried making butter in this way, and churned for seven hours without the desired result. When all my servants, who had churned by turns, were exhausted, I gave in, and sent for a skilful *Gwalla* belonging to a neighbour; he at once asked for a tea-cupful of hot water, and sprinkled it into the jar; before he had given a dozen turns to the churn, the whole surface of the milk (by this time the solid *tire* had been beaten back into a liquid state) was covered with a fine froth of minute particles; a little more sprinkling and gentle churning, and these were beaten into larger grains; then he commenced skimming them out with his hand, and pressed them lightly into balls of about a *chittack* in size. This happened in very cold weather; in hot weather the same process is carried out, but the sprinkling is done with cold water instead of hot. For many reasons I condemn this way of making butter. It is very troublesome, and requires great nicety and skill to do well; the milk is all wasted, as it is very sour and good for nothing, and the butter made in this way is never so fine-flavoured as butter made from cream. The butter-milk is not even fit for the calves to drink, as it will make them ill, and if you give it all to the cows, they will have too much of it.

Susceptibility of Milk

Milk, cream and butter all take the flavour of any strong odour which may be near or about them–milk especially while being boiled. Some smells turn milk sour.

Silver and plated vessels and spoons milk sour.

Iron turns it red, and makes the cream black, but does not make them sour.

Copper, unless well tinned, will also turn milk sour. Brass turns it green, and gives it a nasty flavour.

When it has been kept some time in tin, it turns blue when added to tea.

Fresh earthen pots give the milk, cream and butter a strong and unpleasant earthy taste.

Crockery is a retainer of heat, so that all milk-jugs, cups, and basins are unsuited for keeping milk in this country.

Zinc, copper tinned, bell-metal and wood make the very best milk vessels.

Enamelled ware, being iron, is very cool and nice for milk, butter or cream, and I may say is the best, as it is easily cleaned.

Vanilla has a wonderful effect in keeping milk sweet; a drop of the essence is a great help in keeping it good for tea and coffee.

Cream, Milk, and Butter in the Monsoon

During the rains, milk gives a great deal of trouble, as it is so strongly affected by the constant changes in the atmosphere. Although the weather is cooler, and the thermometer much lower than it was in May and June, the milk will turn; and on very thundery days no amount, even, of boiling will keep it good. Your only plan is to adhere strictly to the second boiling which I have already described. Boil twice during the twelve hours of the day.

This well-boiled skimmed milk makes a very good drink (when cold) for invalids and people suffering from the heat of the weather. A glass of rum or sherry added to it improves the flavour–a milk peg, in fact; in cases of debility it helps the appetite.

Good milk, when sour and just set, is a most wholesome and cooling drink in the hot weather; it should be flavoured with a little wine, nutmeg and sugar; it is very good for people suffering from prickly heat and other disorders caused by the heat of the weather.

BUFFALO'S MILK

The flavour of real buffalo's milk is very fine, but it is known to few Englishmen, who are generally of opinion that the filthy smoky stuff sold in the sweetmeat shops is buffalo's milk; and so it is, but buffalo's milk very much disguised. The innate love of deception which characterises natives of low degree, or rather of telling a bit of the truth, but not the whole truth, with intent to deceive, like school-boys, leads them to make their masters believe that the above-mentioned filthy stuff is pure milk fresh from the buffalo, and that all buffaloes' milk is of the same kind. When a servant is sent in a hurry for milk, he runs to the bazaar and brings his stuff, which is at hand and always obtainable, and if his master complains, it is easy to say that buffalo's milk has been brought by mistake, or that nothing but buffalo's milk was to be procured, thus inferring that buffalo's milk is always bad, and that all bad milk comes from the poor buffalo, the real fact being that the domestic was too lazy to go to the proper place, which was probably at a distance; this would have involved the trouble of having the milk milked into a jug, or taking it from the milking *lota,* which is of bell-metal. Again, if the servant goes for milk at an unseasonable hour, it would be undergoing the smoking process described before, and would be undrinkable. So, all things considered, he finds the easiest and simplest plan is to get milk from a shop, and palm it off on his master as the best buffalo's milk. When the worthy gentleman remembers what an ill-looking creature the buffalo is, he is quite prepared to believe that its milk must be disgusting.

It is not so very many years ago that English people in India declared that the hens of this country always laid stale eggs, and that the milk of the cows was undrinkable; the majority of them never tasted any- thing but bazaar milk, and often longed for the time when they might get home, and once more taste *real* butter and milk. All this is now changed, and every year more attention is paid to the dairy and the poultry yard. The depreciated rupee has no doubt something to say to this; people are not so rich as they were. The fine old Indian Gentleman of the olden time kept a dozen cows and a flock of goats, which he never saw, and for which he had several servants. As long as he got enough milk for his tea, he asked no questions, and those who could not afford to pay so heavy a price for their early tea, went without, or flavoured it with bazaar milk.

Those who keep buffaloes know how useful they are; but they are very expensive, a good one costing eighty rupees in most parts of India, though cheaper ones, giving less milk, are to be obtained in remote districts. Buffaloes are cheap to keep, as they will milk largely when fed on only *bhoosa* and grass, and they are in full milk for twelve months, and go on milking to within a short time of the next calf; they calve every second year. No buffalo is reckoned to be of much value unless she gives from ten to fourteen seers of milk daily. The milk is of a very fine quality, and has a rich, pleasant, sweet taste, and being nourishing and easily digested, it is very good for children. It produces nearly twice as much butter as cow's milk, and the cream from it is the richest and best for whipped creams and sweet dishes. The butter is very nice if made in an English way, but, being white, it should be coloured with saffron seeds, after which it will pass for' cow's butter; and few would know the difference. A home-fed buffalo, fed on gram and *khullie* as well as *bhoosa*, will give excellent milk and butter. This animal is an enormous eater; she thinks nothing of a *maund* of *bhoosa* in the twenty-four hours, so few can afford to feed her on anything but the cheapest fodder. She is also very hardy, and seldom ill or out-of-sorts; she likes mud and marshy places, where she will sit and ruminate for hours, with only her head visible out of the muddy water. The buffalo is not sacred among Hindoos, but a man who possesses one is supposed to be well-to-do, like an Irishman who owns a cow. This creature is the indigenous cow, and is found wild in numbers in the Duars; she certainly stands the climate better than the cow proper.

The bulls are very savage, and in the Duars they often come out of the jungle and join the tame herds belonging to the villagers, much to the, alarm of the herdsmen.

They also produce good sport; a wounded bull sometimes shows fight, and is a dangerous customer.

The tame bull is used for draught and plough. The beef is not badly flavoured, but it has a very dark colour, and is sweetish; when corned, it is very good, and, well pickled, is hardly distinguishable from cow-beef; natives are very fond of the meat. The butter should be dyed with saffron, as should that of cows who have no green food. This is done by rubbing in a few saffron seeds (*suffrun*), which are procurable in the bazaar, and boiling them up with the milk, before it is set; they will not change the look of the milk, but the cream, when it rises, will look yellowish; and the butter, when churned, will come quite yellow, and appear to be of a naturally good colour; it will have a slightly scented flavour not at all unpleasant: and the nice rich colour is so pleasing to the eye that the imagination may be trusted to credit the butter, and not the dye, with the flavour.

In the first edition of this little book, I feel I did not give this subject enough time or space, but knowing how strong Anglo-Indian prejudice is against the buffalo and her milk, I thought it was of little use saying what would not be credited, but lately many people have told me that they cannot make butter equal to my own, even if they follow closely my instructions. The only reason for this, I think, must be that I keep both cows and buffaloes, and mix the cream, for the butter; all dairy farms keep buffaloes, and mix their milk with that of the cows, and make beautiful butter; bought buffaloes' milk in the hills is very good if not mixed with Tree Tallow, which makes it look very rich, and causes a thick cream to rise, and a good quantity of butter also to be produced. It has no flavour, and is evidently quite wholesome. The only way to discover this clever adulteration, is by heating the milk, when brown globules of fat will float on the top, and disappear as soon as it cools. I believe this Tree Tallow adheres to all the pipes in the separators, and gives the dairy men much trouble, but, for all this, good buffalo's milk is to be had in the hills. I am giving some extracts from Mr. Howman's report to Government on Indian milk and dairies, as he speaks so highly of buffalo's milk, and his opinion will, I hope, remove some of the strong prejudice against this milk, which I have found unequalled for cream, butter, cooking, and, after skimming, capital for children. Young children also thrive better on buffalo's milk than on any other, when given, in the usual way, with water and from a bottle. I have known very delicate infants restored to health by taking buffalo's cream and warm water.

After calving, a buffalo gives very little milk for the first few days, and does not give her full amount till the calf is about a month old; they milk on sometimes for over 2 years.

DAIRY INDUSTRIES IN INDIA

Mr. Howman's Report

The Secretary of State for India has forwarded to this country the following report sent to him by Henry A. Howman, Esq. on dairy work in India:

In accordance with your expressed wish that I would communicate to you the result of my enquiries into the dairy work of India, I have to report that I landed in Bombay at the end of October 1889, and returned to England in May 1890, thus spending five months in the country. During that time I visited Bombay, Poona, Bellary, Chennai, Kolkata, Koolna, Bankipore, Nadiad, Cawnpore, Etawah, Saharanpur, Meerut, Delhi, and Allahabad. I regret I was unable to reach Lahore or the Punjab, owing to the lateness of the season. In Bombay I stayed about a month, as it was necessary for me to get practical experience in the composition and treatment of the milk produced by the cows and buffaloes. For the purpose of practically testing these milks I took with me some hand separators of the De Laval make, and I also had a Swede who was competent to work the machines and churn and make butter.

The First Step

My first step was to communicate with the Director of Agriculture for Bombay, who at once instructed some Native officials to attend and give me all the assistance in their power. I may state that I am deeply indebted to the officials throughout India for the assistance they gave me, in every possible way, wherever I went. My first difficulty was to teach the Native what was meant by fresh cream, and to convince them not only that all the *ghi* was contained in the cream that the separator took out of the milk, but also that the separated milk contained all that was required for their use in making curds and curd cheeses, and also sweetmeats, My object was to practically demonstrate this by separating the cream from the milk, making butter from the cream and *ghi* from the butter, then boiling and evaporating the water from the separated milk, thus leaving the solids of milk in a comparatively dry state, out of which the sweetmeats were made. I made special enquiries into the manufacture of sweetmeats, as I found a very large trade was done in this throughout India, providing an article of food much used by the Natives, of which the main constituent was milk, and of which nothing appears to be known in this country.

RICHNESS OF BUFFALO'S MILK

The first point that presented itself to my notice was the great richness of the buffalo milk, so rich in fact that I found a difficulty in passing it through the separator without clogging from its excess of cream. The Government analyst, Dr. Barry, kindly analysed a sample of buffalo's and a sample of cow's milk for me, with the result of showing that there was 7¾ per cent of butter fats in the buffalo's milk, and about 18 per cent of total solids: comparing this with English milk, which gives from 3.5 to 4 per cent of butter fats and from 12 to 13 per cent of total solids, at once shows how much richer it is than the ordinary English milk. The practical test of butter making, after separating, bears out the analysis, for 1 lb. of butter was got from 10.43 lbs. of milk, whereas it takes 30 lbs. of English milk to give 1 lb. of butter. The analysis of the Indian cow's milk shows it is about equal to Jersey milk.

The Richest Milk in the World

The milk of India deserves more than a passing notice, the buffalo milk especially being probably the richest milk in the world, the quantity of butter fat it contains rendering it a great deal too rich for ordinary consumption in a climate like India, where its excess of oil is very likely to act prejudicially on the liver of both children and invalids. In fact, I am inclined to think that some of the illness which is usually attributed to the mixture of impure water with the milk may be caused by its excessive richness; and, in exemplification of this, I have the report of the Scottish Orphanage at Bombay. There the buffalo milk was used; they tried the separator, and gave the children the separated milk to drink. They reported that whereas when they gave the children new milk that it often made them ill, now they drank the separated milk with impunity and relish. The Orphange effected a large saving in expenses, because the children now drank the separated milk instead of tea, and the cream was made into butter saving a fourth of their expenses in tea and sugar. A similar report was given by the European Genera. Hospital at Bombay. I lent them a separator on trial, and they reported of the excellent effect on the patients to whom they gave the separated milk, besides the saving effected by having the cream for making into butter; and I am convinced that if the separators were used in the hospitals throughout India, an immense benefit would result to the patients by the substitution of separated milk as an article of diet instead of using the whole milk.

The present cows and buffaloes are unable to produce milk on the scanty and dry food they are commonly fed with, but, if properly fed with green crops grown in succession, they have every appearance of becoming valuable milking animals. The buffalo clearly, from a butter-making point of view, is an animal of exceptional merit.

BUFFALOES IN KATTIAWAR

A correspondent describing the Kattiawar Agricultural Show which was opened on the 13th instant, writes as follows regarding a herd of buffaloes, which may interest our readers on this side of India. The writer says: "The display of cattle is large, and some buffaloes are here which give forty seers of milk per diem. When one sees what these buffaloes are like, it is easy to understand the large scale on which they give their milk. One measured 8 feet 6 inches around the barrel, and 7 feet 9 inches just behind the shoulder. The horn of a bull measured 2 feet 6 inches in circumference close to the head. These buffaloes are the result of careful breeding and high feeding in the grounds of the, Maharaja of Bhownuggur. His Highness's predecessor paid great attention to the development of the breed and the degree of success achieved is amazing. The heads and horns are bisonic in appearance and size. It was not difficult to believe that two or three of these enormous buffalo-bulls have charged a lion and put him to flight. Curiously enough after hearing this, by-no-means incredible, statement at the show, I saw this evening in an *Illustrated London News*, to hand by the last mail a representation of bisons charging lions. The massive heads and horns over two feet in circumference shown in the illustration accurately represent the armoured heads of these bison-buffaloes of the Bhownuggur Raj."

In buying, care should be taken not to buy a buffalo with an old calf, as has given the best of her milk. I advise buying them before they calve, as then they get accustomed to the new place, and change of food, &c. Many of them are most tiresome about food; the slightest change in their diet, and they will not touch their food; if milking, this will lessen the supply before they take to the new food. In most respects the same treatment is necessary for them as for cows.

They require plenty of water, and quantity, not quality, in the way of food. Great care should be taken when a buffalo "calls," that she should be at once attended to if she is neglected when in season, she will become barren. They call at night. August is their season for calling, and as they calve in 11 or 12 months time, July, August and September are the months for calving,–hence the best months for purchasing. Feed them in the plains on *bhoosa*, and *khulli* (oil cake) and *churrie*; in the hills on Brewery grains and hill grass: quantity, not quality, as said before, is required for these grass feeders,–in fact, gram and grain not only dry up their milk, but give it a rank flavour, so the cheapest food is the best for them. A milking buffalo requires a plentiful supply of salt, but if she is in calf as well as milking, give her salt sparingly and if she is being fed on Brewery grains, something cooling must be added to the grains, such as a pint of butter milk, or a little kuttila gum, or the heating properties of the Brewery grains will cause her to slip her calf.

The dung, mixed with charcoal dust, makes nice balls for burning, and when dry, has no smell, but produces a hot glowing fire, like peat, and very grateful is the warmth of it in the Hills, in the winter months.

The calves are easily reared by being fed on skimmed milk from a common quart beer bottle. While the calf is young, the milk should be warmed, and if it suffers from constipation, a little mustard oil should be given with the milk; if it has looseness, some corn-flour boiled to the milk. A buffalo will not milk without her calf, but it is well to feed it by bottle, as well as allow it what it can get from the mother.

CHURNS

A churn of the pattern given on next page is easily manufactured.

"A" is a simple tin cylinder made any size you may require, according to the amount of milk to be churned.

"B" is a tin lid, lined with a thick piece of wood which fits tightly into cylinder "A"; it has a hole at the top with a tin funnel fastened into it, marked "C."

"D" is a good steel rod with a wooden handle, and at the lower end a circular piece of wood, "E" which has holes cut in it to allow the milk to rush through when it is forced up and down. The lid and rod are one, but the latter must work easily up and down. There is nothing in this home-made churn that cannot be manufactured by any common iron-smith and country carpenter at a small cost. If the churn is a large one, it should have a few bands of iron round the cylinder to prevent the latter bulging, and it should also have a good strong sheet-iron bottom, as servants often churn on uneven ground, which may dent the tin bottom and start a leak.

This churn, though not atmospheric, is on much the same principle, as it is worked by a piston. When the handle is forced down, the circular piece of wood is pressed down on the milk, which bubbles up through the openings; and when the handle is drawn up, the milk rushes down again, and these motions continued, beat it up into a foam. Years ago, in the days of our grandfathers, churning was done by the upright motion, but for many years past a rotatory

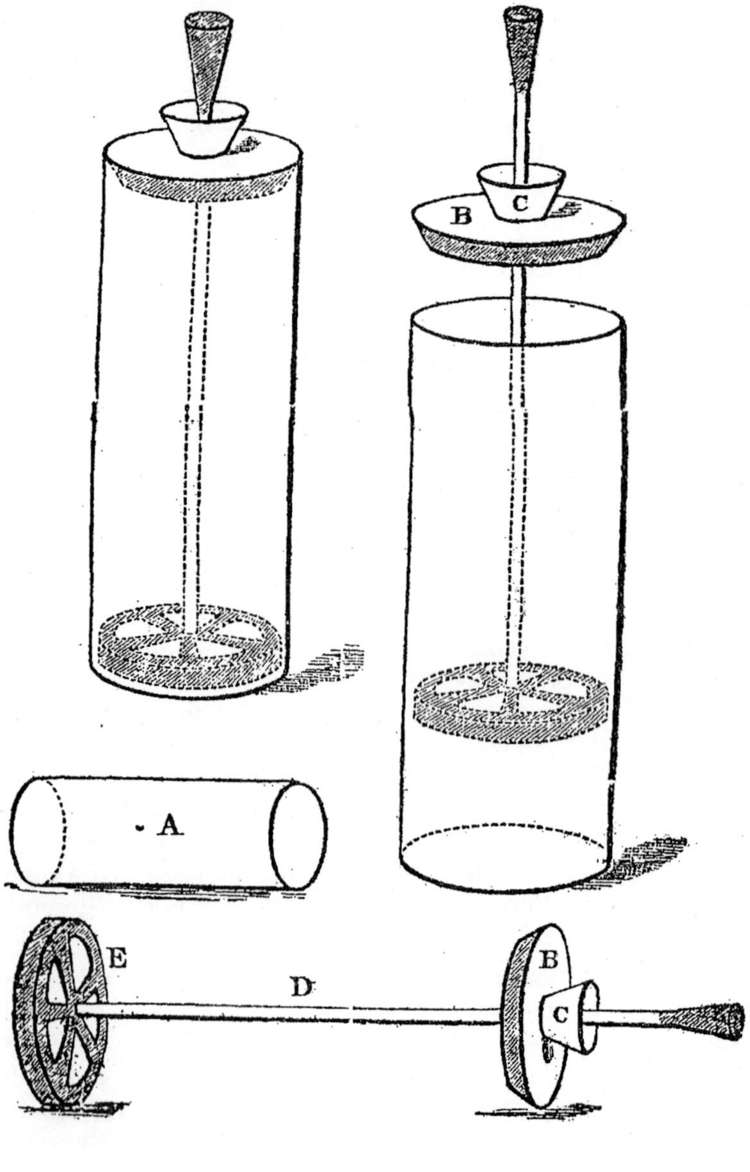
B
C
A
E
D
B
C

motion has been adopted, which is once more being replaced by the original up-and-down action, and there is little doubt that this lifting and lowering action is the most successful, as it is more in accordance with scientific principles. The millwheel plan, or fly-wings of wood going round, was good, but there was so much leakage at the handle that it was messy; and in the rotating plan, where wings turned round in a kind of bucket, the motion was too smooth, and did not force out the butter quickly enough. I think the up-and-down action is now firmly established in all practical dairies. Atmospheric churns are the best patent churns yet invented, and in churning milk they are unrivalled.

SEPARATORS

Here I give a full description of a Separator and Barrel-Churn which will be interesting to my readers, though both the above are of little or no use in private families, for they only make good butter in large quantities.

The machines used are Laval's Patent Cream Separators, and as the general system of production is little known in these parts, a short description of it may not be without interest to our readers. The buffalo milk (which is employed by reason of its containing more cream than the ordinary cow's milk in this country) is first put into a small reservoir placed on the top of the machine, and escapes by means of a tap through a pipe fitted with a float to regulate the supply, into a cylindrical receptacle in which the work of separation is performed. The interior of the cylinder, resembling a bottle in shape, contains two fixed metal wings which keep the milk in motion while the cylinder is revolving, and some idea of the rate at which the latter revolves may be gathered from the fact that forty turns of the handle–all the machinery being worked by hand–produce no fewer than 6,000 revolutions of the cylinder per minute. Thus by means of centrifugal force is the milk separated from the cream in an incredibly short space of time, the purest Swedish steel, of which the cylinder is made, being more than equal to the tremendous strain occasioned by the high velocity of the separator. At the far end of the cylinder, and placed in close proximity to its circumference, is a little tube curved in an opposite direction to that in which the separator revolves, so that the milk, which is the heavier body, rushes into this and is discharged, through a small hole pierced in the tube, into a tin placed beneath to receive it, while the cream, which is lighter, passes out by another tube into a second receptacle. All the machines are constructed on the same principle, the only difference between the two in use at the dairy in Tamarind Lane being that in the larger one, known as the Windsor Separator, the cylinder is placed in a horizontal position, whereas the cylinder in the other, the Baby Separator, is vertical, while of course there is, owing to the variation in size, a considerable difference in the matter of capacity. For instance, the former is capable of separating 36 gallons of milk per hour, while capacity of the latter is only 13 gallons an hour. One-sixth part of the contents of the cylinder in each case is cream, and five-sixths skimmed milk, the separated milk being discharged from the machine in less than two minutes from the time the operation commences, and the cream about half a minute later. After the cream is separated, it is allowed to stand for something like twelve hours in order that it may ripen, and then it is placed in the Victoria Churn, a barrel-churn, which is easily turned by a small handle, and in about fifteen minutes the butter is produced, one of the advantages of the Victoria Churn being, that it is not encumbered with beaters, therefore the process of cleaning is considerably facilitated. The percentage of butter obtained from the cream put in the churn is about one-third part of the entire contents. The butter in its granular state is taken out of the churn and placed on a board, where the water which has been used in washing it is extracted by means of crinkled rollers, which are passed over it several times. Then it is weighed and made up into small pats, ready for the consumer.

NATIVE CHURNS

The native plan of churning is simple, but not very effective. The apparatus consists of a long stick, to the bottom of which is attached a wooden cross, a large earthen jar or pot, and three leather straps. The milk is put into the jar, which is kept in an upright position by stones; the wooden cross is placed in the jar, two of the leather straps attach-the stick to a tree or upright pole, and the third strap is passed round the stick two or three times, the ends being retained in the two hands. By pulling this strap alternately with each hand, a rapid rotatory motion is given to the cross, which works up the milk until it is churned.

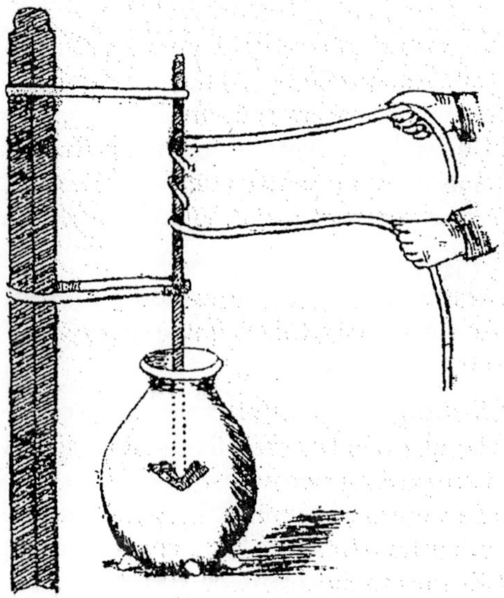

DURATION OF MILKING

Opinions differ both as to the time cows remain in calf, and also as to the duration of milking. Some say that a cow should milk up to a month of calving, but I have never met with one who did so. My experience is that a cow is in calf for about 290 days, *i.e.*, a little over the ten lunar months. Age and condition of health make a difference.

Cows are in full milk for four months, on an average; from four to six they show signs of lessening, by small degrees; during the seventh month there will be a decided difference, and about this time they again want to calve; they go on milking for two, three or four months longer, and at the end of the time will still give about half a seer of milk.

When a cow is going off her milk, any change of food will often accelerate the decrease, and she will not recover herself, but the milk will stop altogether. When a cow seems to be going off her milk without a cause, it may be that she is out of sorts and that her digestion is out of order, in which case an ounce of dry Chiretta sticks pounded and made into a ball, and given for a few days, will often restore her appetite, and give her digestion the tone it requires.

CALVING

If a cow in full milk with a young calf should give signs of wanting a bull, she should be fed on butter-milk for a few days, as it is evident that her desires are not natural, but result from overfeeding; and, if gratified, will result in drying her milk, which, while it lasts, will be unwholesome for young children. The milk of a cow carrying a calf turns sour very soon, and, when boiled, sometimes becomes curds and whey.

The average run of good cows should have, roughly speaking, two calves in three years; that is, seven or, eight months in milk and ten months in calf.

If rich feeding has the effect of making a cow want to "go out," she should be given butter-milk the first thing in the morning for a few days disregarding her wants will do her no harm but should her calf be of an age to graze with ease, when the mother will naturally begin to dry, it is time for her to begin producing a new calf, and *then* it is very injurious to the cow to neglect the wants of nature, and if she is so neglected, she may become barren, or slip her calf next time, which is a sign of weakness.

When the time has arrived for a cow to get into calf, she shows signs of her state by calling, and sometimes stamping restlessly with her fore feet, and pulling at her rope. She keeps her tail lifted up. but she shows no other signs. She should be sent out at once, as time is an object, this state lasting only an hour or so; if the time passes by, the opportunity will not again occur for perhaps months. Some cows become very troublesome, even savage, and will butt; others show their state but little, often doing no more than lowing gently; these latter are the cows most generally neglected, as their state is not observed. A very fat, overfed cow will want to go out time after time before she is in calf. March and April are the months when nearly all cows want to go out, so a barren cow should be carefully observed during these months, for if she calves once, she may turn over a new leaf, and take to calving again regularly.

If a cow remains more than a year dry, and yet is not in calf, it is as well to send her out altogether, to remain with other cattle, grazing and feeding with them; the company and the change of air and food may have the desired result. It is very seldom necessary to help nature in this matter, but if it is, spices, *ghee*, *goor*, ginger and mutton suet, will all help to bring a cow into calving condition. Heifers generally want to go out in the spring of their fourth year (but if in poor condition, in the autumn of their fourth year), while they are changing their teeth; a good rule to go by is that the heifer will have her first calf at the same time that her mother has her third calf. A cow that calves too often is as bad as one that calves too seldom, as if she has a young one once a year, she remains a very short time on full milk, and also breaks down sooner, and comes to an end of her calving by being in too great a hurry over it.

If a cow is troublesome about getting into calf, give her the refuse liquor from a native distillery; it is the juice of the *mahawa,* and will often answer the purpose. If given to a cow in milk, it will dry her. as it is of a very heating nature.

In the Hills cows often slip their calves from eating the fallen pine-leaves, and for the same reason it is dangerous to give them pine-leaf bedding, as they sometimes eat it during the night.

Dry cows in calf should not have salt.

If a cow gets into a heated state of blood when in calf, give her early every morning either butter, milk or *kutteela* gum, soaked in cold water over night.

BULLS

Every cow has a different constitution, so it is impossible to lay down any rule for their milking powers; no two cows milk the same time, or give the same quantity, hence the great difference in their breed and value; two sister heifers will not prove alike; sometimes the daughter of a splendid cow will be a very poor milker, and *vice versa;* but great care should be taken to select a good bull.

A good bull of a fine milking breed will generally have good milking daughters. This is a subject to which sufficient attention is not always paid. Bulls are chose more for the purpose of improving the bull-calves that the cow-calves. It is a mistake to send your cows out into the main ruck of cattle. Nearly every station in India has Government bulls, which are to be had for the asking, and if your cows are served by them you insure the production of havored calves. Government bulls are not always successful, as they are often overused, and oftenest overfed and too fat; in these cases the only plan is to go on sending the cows until the end is gained. As the cowkeeper finds this troublesome, he may take the cow to the first bull which is at hand, which may be an animal of no breed at all; this must be guarded against. This indolence on the part of servants is the cause of many indifferent calves begin born, to the astonishment and disappointment of the owner. It is a peculiar fact that when a cow has been served by a well-bred bull and has a good calf, she will milk better than when she has been served by an ill-bred bull and has an indifferent calf, particularly in the case of cow-calves. I can give no reason for this, but state the fact as the result of my observation. Perhaps nature supplies more milk when the calf is a large one, just as a goat will give more milk when she has two kids than she will when she has only one kid. Heifers are sometimes very troublesome on account of their timidity, and, for this reason, frequently do not get into calf till long after they are old enough, and have to be sent out many times before a satisfactory result is arrived at. When the cow or heifer comes home, she should be tied up and left undisturbed; she will generally sit down quietly, and not more for hours. It is as well not to give her much food that day; grass and water are quite enough, and if she is a cow given to disappointing let her have a feed of *oorad dall*, a little *kuteela* gum soaked. If the cow was in milk when she was sent out, the milk will decrease rapidly, altogether will produce just as much butter, if not more, for the quantity of milk.

As a rule, cow-calves take bull-calves after the mother.

DRY COWS

It is difficult to decide satisfactorily when to reduce a cow's feed from full feed to the spare diet of a dry cow. When a cow is going off her milk and is also in calf, stopping her food will naturally drive away altogether the supply which is already inclined to cease; so it is as well to keep up her supply of food to suit your own convenience. If you have no other cow in milk, or require her milk for butter, continue to feed her up: but if you have sufficient milk from other cows, or do not think the quantity she gives worth the expense of feeding her, it is better to stop doing so, and leave her calf loose to drink what she gives; it strengthens the calf and is good for the cow, as she allows the calf to suck or not, according to her own feelings.

Dry cows should not have grain and good food, which only fatten them, and fat in a cow is serious defect as it makes her liable to slipping her calf, and also becoming barren; it is a common cause of failure with the cows of the *sahib log*. Observe the half-starved native cows, who go on calving time after time with regularity, although they do not appear to have sufficient strength to eke out their own poor lives, much less to produce others. A dry cow should have plenty of grazing during the day, and a little cat grass or *bhoosa* to munch during the night, and she should be decidedly thin to within a short time of calving, when she will pick up of herself if she gets a sufficient quantity of grass. In the rains, and wherever grass is plentiful, grazing during the day is all that it necessary; nothing need be given at night. Overfeeding when a cow is not in milk makes the food run to fat, and if a cow's food once begins to do this, it will continue to do so when she is fed up after calving, and in a short time she will become fit only for Christmas beef; though she may have a large udder and a huge belly, her milk will be scanty. But if the good food comes at calving with the fresh flow of milk, all the food goes to the milk, as the system has become accustomed to spare diet and, being healthy, her muscles will improve, but she will not put on flesh. For the same reason she should have plenty of exercise, which is very necessary for a cow. If she is allowed to get fat before calving her calf will be born puny, all the nourishment having been monopolised by the mother.

In the early months it is difficult to decide if a cow is or is not in calf, and it is then that it is most necessary to know, both to prevent her lying fallow when she ought to be in calf, and also that there may be no mistake when buying and selling. The most certain sign of a cow carrying is a discharge, however small. In the absence of this, it is probable that she is not in calf. I remember the case of a cow showing *no signs whatever*, except this one, up to the very day she surprised her keeper by presenting him with a calf: she was supposed to be barren. Generally, in the latter months, a cow's condition is quite evident from her size.

PAPEETA LEAVES AND FRUIT

Papeeta leaves and fruit will often bring back a flow of milk if it has stopped from any disorder in the cow's health. The leaves must be pounded and given as balls. The fruit is sweet, so readily eaten.

MANAGEMENT OF COWS WHEN CALVING

When a cow is within a few days of calving, her shape will undergo a change; she will become hollow just beneath her hip-bones, and her stomach will appear as if it had sunk down towards her chest; in the case of an old cow, it is very evident that a great change has taken place in the position of the calf. As soon as this change is observed, the cow should no longer be sent out to graze, but should be kept at home, in order that the event may not be unduly hurried from any unforeseen cause, but may be allowed to take its natural course. If a cow calve away from home, she runs a chance of catching a chill. Some cows calve very quickly, even the very day the change is observed, while others may not do so for a week. I am sure keeping them quiet has often something to do with it. About ten days before calving, the udder increases, and sometimes becomes full of milk, and the milk-vein is distended. An hour or two before the event, the face of the cow will have an anxious look, the eyes will be bright and staring, and hollow, indicative of pain. As soon as the cow shows these signs, she should be taken into her house and kept quiet; some straw should be spread on the floor, and she should be given a good supply of grass to amuse her and keep her from fretting. Her keeper should remain close at hand, but be careful not to disturb her unnecessarily, nor sit and watch her. She will eat a little at times when out of pain, When she begins to sit down and get up, as if very uneasy, the man should stay with her until the calf is born, and afterwards he should prevent her from getting up until the calf is quite clear of its mother; but, at this stage, nearly all natives are good nurses, and know what to do. Where they fail is in the management before and after birth. When the cow and calf have been attended to, and there is time to send to the bazaar, a quarter of a *seer* (or half a pound) of *goor* and the same quantity of ginger should be obtained, together with enough *atta* to make them up into balls, and the cow should be given half the mixture at once, as it helps to clear away the after-birth, and to reduce the after-pains; which trouble and weaken some well-bred cows very much; many of the best milkers are very delicate, and suffer much at the time of calving. The remainder of the balls should be given about six hours afterwards, and then no more of this mixture should be given, as it would be hurtful as soon as the milk flows. Care must be taken that the cow has nothing to drink, and that she has a warm coat thrown over her, as she is very apt to catch cold during the first few hours succeeding the birth of her calf.

If she is doing well the less you disturb her the better; but if her eyes become glassy, and appear of a steel colour when the light is thrown on them, and the hollows above them deepen, she is in pain, and about four hours after the birth she should have a hot bran mash, rather sloppy, which will fill and warm her inside, and quell the pain; this should be continued for the first two or three days, as it assists the milk to come, keeps her inside warm, and opens the bowels. For the first day she should have no water, and then warm water for a week; this is a point to be very careful about, as a draught of cold water will chill her milk, and may produce swollen udder and inflamed nipples–a most tiresome complaint resulting in the cow drying up–as she suffers so much pain that she will not allow herself to be milked, nor her calf to suck.

It must be remembered that cows who are good milkers are unnatural cows; they are like cultivated plants, very delicate; a common cow who gives only enough milk for her young one, will stand any amount of neglect, but a good milker will go wrong in no time.

A common complaint with large, loose-limbed cows–especially if they are old and have had many calves–is a slipping down of the womb; this is a complaint that natives do not know how to treat, and they generally let the sufferer die; and, if a cow is known to be liable to it, they will not buy her at any price, however low.

But it really only wants care. As soon as it slips it should be carefully washed with luke-warm water and alum, and then replaced by a skilful native, and an alum and water enema given; the cow must not be allowed to sit down, and she must be watched, and not allowed to strain when in pain; if she does so, or if her eyes appear glassy, she should have a dose of bazaar opium–about one anna's worth at a time-dissolved in water.

About an hour or so after a cow has calved, she should be milked (about this milk the natives are very freedy, as they say it is as rich as cream, and with it they make some very dainty dishes), and the calf set to suck; the cow should be milked three times a day, every drop of milk being extracted, and then the calf should be allowed to mouth and suck for an hour at a time, as the sucking opens the mill, vessels, and brings the milking machinery into play; the calf, being hungry, will suck hard, and draw off pain from the cow. On the third day, the real flow of milk will set in, the udder will increase, and the milk become of a better flour; after the third day the milk is quite fit to use, though still of a yellowish tinge; and if butter be made from it, there will be a large quantity of a rich colour and flavour. If this new milk, while still yellow, is given to a baby, it has a purgative effect. As soon as the full flow of milk sets in, it will become lighter in colour; the milk of a cow for the first month is poor, and produces less butter than afterwards–hence it is good for young babies, or for those who have been lately weaned; when turned, it forms small curds. The milk of a young cow with a young calf forms feathery curds, while the milk of an old cow, and of one who has been in milk long, forms larger curds, and that of a cow who is carrying another calf, curds with very large grains, forming lumps almost as large as nuts; this may be the reason why such milk produces pain during digestion by young children. Mothers, when buying a cow for their babies, should ascertain the age both of cow and calf.

After the first week, the cow should be milked. morning and evening, and always at the same hours, as then the flow of milk comes at the time it is wanted.

A Good Mixture for a Weakly Cow Immediately After Calving: One seer of *Ruhur Dall*, boiled; one chittack of salt; one chittack of *Haldee*, or Turmeric, pounded. Half to be given in the morning and half in the evening.

MILKING

A cow should never be milked before feeding; after she has finished her food, the calf should be brought out. No native can milk a cow without first allowing the calf to mouth the nipples, as this opens the valves, and the cow, seeing her calf near her allows the milk to flow; sometimes you will see both udder and milk increase rapidly in the space of a few minutes. The cow will now relieve herself as nature dictates, and, when she has done so, her calf should be tied to a peg near her head, where she can lick and fondle it while she is being milked. A kicking cow should be hobbled, but not a good-tempered one, as it is a had babit to get them into, and may teach them to kick. Let the cow be always milked by the same man, for a change of milkers may make her restive, and cows like being milked, if the operator has a gentle touch. In fact, cows are creatures of habit, loving peace and quiet and regular routine; if they are always fed at the same hour, and milked at the same hour, in the same place and by the same man, they will seldom rebel.

If cows are troublesome at milking, get rid of the *milker*, as a good man will have his cows quite docile; and though one may be ticklish and difficult to milk when first put under his care, in a short time he will have her quiet and patient. If a cow takes a fancy into her head not to be milked, and you are sure it is not the fault of the milker, and that he has neither tickled nor pinched her, and she does not want to calve again, look to her nipples; they are probably chapped, and as she cannot speak, and the pain of being milked is very great, she kicks about to let you know her distress.

Chapped nipples are most tiresome, and with some cows they are constitutional, as sand crack with some horses; but, as a rule, they are the result of leaving the nipples wet after milking, and allowing them to dry in the open air. To avoid them, the cow's udder should be well cared for, and, from the very first, butter should be rubbed in, or a mixture of wax and fat and candle-grease (equal parts) boiled together into a soft ointment. It is well to rub some greasy matter over the nipples for a good month after calving, as then they harden without chapping. Chapped nipples are very painful, and may dry the cow. If the calf dips its mother with its teeth, it must not be left to mouth for too long a time.

After the milker has drawn away all that he can get, the calf should be allowed to suck again, so that every drop may be extracted; any milk left helps to dry the cow, as nature will absorb everything that is extra, and the next flow will be lessened. The more you milk and the more the calf sucks during the first fortnight, the more milk will be produced. In the event of a cow's milk lessening suddenly, it is as well to let her calf loose for a day or two, as its continual sucking may bring back the milk, which has had a check.

Natives cannot milk cows in the manner Englishmen do, and unless the calf sucks first, and brings down the milk, they have great difficulty in extracting anything. An English milker grasps the udder with his whole hand, and drives the milk out by strong pressure on the milk-valve above the teat; the native milker holds the teat only with finger and thumb, and drags out what he can; consequently, if the calf dies, there is great trouble, and the poor dead little thing has to be stuffed, and stuck up before its mother for her to lick and look at while the milking goes on–unless this is done she will not let her milk down; but even with this device, the milk will gradually lessen, as the native will not be able to extract every drop. As soon as a calf dies, send for a *mochee.* or leather-tanner and have the skin cured and stuffed, and dried in the sun.

When a cow calves before her time, the calf gene rally dies, or is born blind, or something else goes wrong; if it is going to die, it will probably do so during the first month, or about the time it should have been born. Care and management will save the cow's milk, but she will not milk so well as she would have done if her calf had been hearty and had lived. Heating food, over-feeding, bad cough, too frequent calving, over-fed bulls, over-used bulls sudden and long marches, blows from sticks and other injuries will cause premature birth. In some cases the fault lies with the cow herself, who is inclined to be barren, and will, after raising her owner's hopes, fail him by calving a month or six weeks before her time.

In the Hills this disappointing event is common. There are many heating herbs, as well as the fallen pine leaves, which help to cause this disaster, and if the cows are loose, they often slip up and down the hill-side, which in their heavy state injures them, and makes them calve before their time.

Sometimes they will calve a few days too soon, and in this case the calf will be weak, and will not suck, and generally dies; if the calf dies, it is as well to blindfold the cow, borrow a calf from another cow, and let it suck, to get the milk flowing freely, and if the cow objects, her dead, stuffed calf should be placed before her nose; this will deceive the poor cow, and she will allow the borrowed calf to suck.

Sudden swelling of the udder is a serious symptom, and if not taken in hand at once, may lead to much trouble, drying the cow, and perhaps preventing her from ever milking again. It is a tendency to gathering, caused by a chill, or by a bruise, or, in many cases, by too high feeding before calving. If the udder gathers, the nipple becomes what the natives call blind, *i.e.*, the milk-valve closes, and the vessels thicken and choke up, so that that side of the udder becomes useless flesh or fatty substance. As soon as it is observed that the udder is swelling, and that there is difficulty in extracting the milk, *neem* leaves should be boiled, and the pot put under the cow, so that the steam may pass over the affected, parts, which should also be well rubbed with mustard-seed oil, and the calf should be set to suck the nipple just under the swelling in order to draw off some of the inflammation. If the cow be very fat, give her a very large dose of castor oil, and keep up the fomenting; a few hours' neglect at this time may ruin the cow–if the udder once gathers, it is almost certain that she will lose the use of that part of it–perhaps of the whole udder. If a cow has a blind nipple, or there is any fear of a troublesome udder, let her have a couple of doses of castor oil on the day of the flow of milk, *i.e.*, the third day after calving; also rub her udder with warm mustard-seed oil, but do not foment unless there are signs of inflammation and swelling; one of the first signs is the cow objecting to the bumping of her calf's nose–this shows that painful lumps are forming, which will not bear pressure; a plaster made of pounded turmeric (*haldee*) and lime, well spread all over the swollen part of the udder, will help to set the circulation going, and absorb the lumps.

COMPLAINTS

Cows well fed and cleanly kept are subject to few ills, except disorders produced by over-feeding and want of exercise, and chills; they are naturally very hardy, and liable to few illnessess, so simple bazar medicines and a little knowledge regarding their complaints will cure most of their indispositions.

Scarred Cows

Oedematous scars are sometimes to be seen about the throats and necks of cows, particularly those from Central India. These scars look like a kind of cow cancer; they do not kill, and are perfectly harmless as far as the milk is concerned; some of the largest milkers have them. Of course a cow is better without them, but their presence is not sufficient reason for rejecting a cow. The native cure for a sore of this kind is firing, but it will often break out again, and the cow will remain in good health as long as there is a slight discharge from the sore; but if the discharge is checked, the cow goes off her feed, and is generally out of sorts; the part in which the sore is situated will swell, and a running at the mouth, nose, and eyes, sometimes accompanied by looseness of the bowels, will set in. The scar must be opened, and the cow will be relieved at once, and will be restored to her normal health. Natives call this complaint *batauree* or blind boil, and consider it no blemish. It is not infectious, but the calves may inherit it sometimes, on its first appearance, they become very ill, but it seldom affects their health afterwards.

Cough

Give bamboo leaves every morning. If the cow will not eat the leaves, pound and give as a pill. Some cows cough all the time they are in calf, and should not be dosed, as the cough is of no consequence.

Hypertrophied Papillae of the Mouth

Natives call this complaint *kanta*; it is a curious complaint, and is caused by a disordered stomach. The mouth, cheeks, lips, and tongue become so tender that the cow cannot eat. All ruminating animals have pointed *papillae* like points or thorns on their tongues, and in the lining of their cheeks. When the complaint we are discussing attacks a cow, these *papillae* grow long, and have sharp hard points, sometimes an inch long and as hard as steel; when the cow attempts to eat, they hurt her so much that she is forced to give up food and starves, and, of course in the case of a milking cow, her milk rapidly disappears. At the same time her mouth becomes yellow and furred, and her breath bad. When a cow's breath is disagreeable, you may be sure that she is very much out of health. The best way to get rid of these sharp points is to send for a native barber, have the cow thrown, and the points shaved off with a razor, after which the mouth should be well rubbed with salt, and the cow be given a purgative dose; she should be forced to take a large quantity of salt, daily for about a week. Cows who are regularly supplied with sufficient salt, and have plenty of bran in the food seldom are troubled with this complaint.

Weak Eyes

Wash with warm water, and then blow a little sulphate of zinc into the open eye; use a tube made with stiff note-paper rolled.

Diarrhoea

Looseness of the bowels is a most tiresome complaint, as it weakens a cow and prevents her giving as much milk as she otherwise would do. Try alum in the drinking water; if this not efficacious, give common chalk twice a day in *atta*; if the looseness is still unchecked, pound up unripe pomegranate fruit and make into pills, about four pills to each fruit, and give one pill morning and evening.

Worms

When the looseness is caused by worms, other remedies must be resorted to. The best thing is a dose of *til oil* with turpentine, one dessert spoon of the latter to a pint of the former.

Weak Digestion

Most frequently loose bowels are the result of weak digestion; good feeding while milking keeping up the irritation; salt and bran should be stopped, and *bel* leaves given in sufficient quantity.

In many cases the looseness returns when the remedies are stopped, so it is a great fault in a cow, almost enough to condemn her; and, unfortunately, it is often the best-bred animals and the largest milkers who are subject, to it; when obstinate, it will cause barrenness, or make a cow liable to slipping her calf. Natives believe that an untimely calf is always blind; this I cannot vouch for, though it certainly was the case that a calf of mine, prematurely born, was blind, and was deformed in no other way.

Recipes for a Purgative

The following were given me by a gentleman who keeps a large number of cows in the Hills, and has a flourishing dairy:

1. Epsom Salts	8 ozs.
Sulphur	4 ozs.
Ginger	2 drms.
Warm Water	1 pint
Linseed Oil	12 ozs.
2. Epsom Salts	6 or 8 ozs.
Castor Oil	8 ozs.
Gruel	1 ½ pint
Ginger	½ oz.

This purgative is most useful in the case of a cow being poisoned, and should be followed for many days by doses of castor-oil and opium, to soothe the bowels and keep them open, and the cow fed on *suttoo* or *atta* Porridge. *Gwallas* and *chamars* often poison cows with arsenic.

Swelled Neck in Calves

A very dangerous complaint is a swelling in the glands about the throat and gullet of a calf. It is very catching and comes on quite suddenly, and apparently from no other cause than that. It is prevalent, much as mumps appear among children. The first sign is a bad cold; then the coat grows rough and thin, swellings appear in the glands, and the calf will not suck. In a day or two–sometimes in a few hours only–the teeth set the patient seems unable to move its jaws, the eyes become glassy, and the calf dies. Occasionally, if prompt remedies are administered in an early stage, it recovers, but is weakly and sickly for months afterwards. A dose of castor oil should be. given (two large table-spoonfuls) and fomentations and bran poultices applied to the swellings in the neck. If the calf recovers, it should have tonics and strengthening food, and great care for a long time afterwards. As soon as the disease appears, the calf should be isolated; all vessels it has used should be destroyed, and its house thoroughly cleaned, straw and grass burnt, and the walls plastered with fresh mud, as the complaint is most infectious, but only calves suffer from it.

Sores in Feet and Mouth

If a cow has a sore mouth or sore feet, it does not follow that it has foot and mouth disease, but these symptoms are suspicious, and it is always well to isolate her and *not* drink her milk until you are sure that she is not diseased. Cattle often get very nasty sores at the division of the cloven hoof from being allowed to run about with a long rope attached to their horns or necks, which trails along the ground; the poor beast steps on it, and it runs through the cleft in the hoof and cuts the flesh. At first it is nothing more than a rope-gall, but flies get to it, and it becomes a nasty sloughing sore. In this case a poultice made of pounded custard apple-leaves (*Sureefa kee Puttee*), or a bran poultice with a few drops of turpentine sprinkled on it, will destroy the maggots, and when they have disappeared, the sore should be well washed with warm water and alum, and then greased over with vaseline or glycerine. This should be repeated for some days.

A cow's mouth sometimes becomes very sore for want of salt or green food, in which case give her plenty of salt and green cabbage leaves; or it may be that her stomach is disordered, when her mouth will be yellow, and her breath disagreeable; then give her a purgative and plenty of salt, and follow up with a chiretta tonic until her health is restored.

Lampas is a swelling in the roof of the mouth, about the tongue, and close to where the front teeth press, and prevents the cow from eating. The mouth should he rubbed with salt, and if that does not reduce the swelling, it should be scored with a hot iron, and rubbed with burnt paper or rag which is hot from the fire.

Foot and Mouth Disease

If a cow really has foot and mouth disease, the safest plan is to take the advice of a veterinary surgeon; if none be at hand, go on with the chiretta tonic, put her on soft food, and keep her feet in very wet bran poultices, which must be put into leather boots. Care and time will often cure mild cases; where many cattle are affected, sufficient care cannot always be bestowed on them, and many die in consequence from neglect. It is a long and tiresome complaint, as the whole animal seems to become diseased for a time, and the greatest care and management and patience are required, for there is often a relapse, even after the cow has apparently been cured.

Natives have a very simple and practical way of easing the sore feet; they tie the cow in a bog, so that they are thoroughly soaked; this in many cases cures the breaking out at the feet.

The plan adopted by the Commissariat is painting the sore places with Venetian Turpentine, known in the Bazaar as *Gunda Baroza*. A strong solution of Phenyle is good too.

The greatest care must be taken that the affected beast does not graze with other cattle, as the saliva will carry the disease to others; grass trampled by the sore feet will also give the complaint, as will drinking water; there is a running at the mouth, nose and eyes.

There is not much to be done for the mouth except fomentation with boiled *neem* leaves. The treatment consists in restoring the general health, and assisting the sores to discharge, as the whole system appears to be poisoned, and must be assisted to get rid of the poison. If the case is a very obstinate or severe one, it is better to destroy the animal, as it takes so many months to restore it to health. Fortunately it is not a disease which often attacks well-bred cows; it is the common country cattle which are subject to it.

Chiretta Tonics

Simple

Pound up chiretta sticks, procurable in every bazaar, with a little *Atta*, and make into balls, each ball to contain ½ oz. of chiretta and 4 oz. of *Atta*.

Strong

Soak 4 ozs. of Chiretta in a bottle of Rum. Dose–a small tumblerful put into a quart bottle and filled up with warm water. Chiretta tonics is one of the most useful medicines for cows. It is very cheap, and is a cure for most of thè milder complaints to which cows are subject, as it has a wonderful power of restoring both stomach and blood to a healthy state.

To Dose a Cow

Make her sit down by the following method: Tie up a fore-leg, then fix a rope to her tail, pass it between her hind legs, and pull; she will sit down at once. A man should then sit on her neck, and raise her head by holding her nostrils with one hand; thrust the other hand into her mouth, seize her tongue, and draw it to one side of her mouth, while another man passes the bottle down her throat by the side of her tongue, and gives her the dose.

Cow-pox

Cow-pox will sometimes make its appearance on the cow's udder, often only a pock or two, in which case it is of no consequence; but should the udder swell, and large pocks make their appearance, much care is necessary, or the milk will dry up. The cow will not allow her calf to suck, and becomes troublesome to milk, as the udder is tender. The best plan is to steam the udder by holding a basin of hot water under it, or, better still, to tie the cow's hind legs and wash the udder with tepid water and a sponge of soft cloth. She will like this; as she gets accustomed to the heat of the water, increase it from luke-warm to warm, but not too hot on any account, as hot water will boil the skin of the udder and make it doubly tender. When the udder feels soft, wipe it with a soft cloth, and rub it over with butter or *ghee*. Then commence milking, being careful to avoid pressing the fingers on the actual pocks. Thus, with care and perseverance extract all the milk, or as much as you can, and continue to do this daily; the disease does not last long, and in no way hurts the milk, except by drying it up if it is neglected, and left in the swollen udder. Keep up the greasing for some time after the pocks begin to heal, but stop the warm water bathing as soon as you can get out the milk without its help. It is a very catching disorder, and the man who milks the affected cow should not go near other cows in milk. When not in milk it matters little if a cow has it or not; in fact, it is hardly observable, and in all other respects the animal is perfectly healthy, eating and drinking as usual, and with a soft, black moist nose, as if she were in rude health.

I have observed in cases where the cow-pox is very bad that the cow has small lumps all over her and under the skin, but so soft as not to be felt by the hand, but they give the coat a rough lumpy look only perceptible to an experienced eye.

It is from these pustules in cow-pox the lymph is taken for vaccination.

There is another kind of cow-pox which in many cows is always breaking out. The pustules are small and dry and round, only occasionally rising enough to produce a bleb of pus, and seeming to give the cow no inconvenience, unless the pustules form actually on the mouth of the milk-valve. In real cow-pox each pock is as large as a four-anna bit, and exactly like the vaccination on a child's arm, going through similar stages, and taking the same number of days to run its course.

Colic

If a cow is supposed to have eaten something unwholesome, give her a large dose of castor oil (*rairee ka til*) or linseed oil, followed up after a couple of hours by a bottle of strong, hot tea. This will probably be sufficient, but if not, give the following dose:

Colic Ball

Cubebs (*Kubaba*)	4 drams.
Bishop Seed (*Ujwain*)	2 drams.
Red Pepper (*Lal Mirch*)	1 dram.
Opium (*Opheem*)	20 grains.
Camphor (*Kaphoor*)	20 grains.
Venice Turpentine (*Ganda Beroza*)	2 drams.
Calomel	20 grains.

Mix into a Ball

A man should be set to rub on each side, over the place where the pain is, unless the cow is large with calf, when rubbing must not be resorted to.

Cows seldom get colic, but half-grown bulls often do.

Gripes are generally produced by cold, or by bad food. If fowls are allowed to scratch about near cows, their droppings are pretty certain to produce mild attacks, in which case a dose of castor oil is quite sufficient.,

A FEW GENERAL REMARKS ON THE BEST KNOWN INDIAN BREEDS OF COWS

Hansi or Hissar cows are thought a great deal of as they are very large and handsome. They have broad heads, short necks, high humps, square hind quarters, many of them looking more like bullocks for draught than cows.

They average in price from 50 to 80 rupees each.

The milk is of the best quality, and the colour generally of a creamy white. In their own district they give from ten to fourteen seers of milk, but when removed to other parts of India, they rarely give more than eight or nine–sometimes ten. They miss the beautiful grass of their own part; this grass is half the secret of the superiority of these cattle.

They are very delicate, and not prolific.

A half-bred Hansi cow is often very satisfactory, as she will combine the good qualities of both breeds and is not so delicate as the pure-bred animal.

The Nagourie cattle are of the well-known trotting breed used by Hindoo gentlemen for their covered bullock-carriages. They are much prized, and, if good trotters, fetch from two to four hundred rupees each. Before the Indian Mutiny they were greatly used in all large cities by rich native gentlemen, and, in those days, were carefully bred for the purpose. Now they are not so well preserved, and you only occasionally meet with a really well-bred pair of trotting bullocks. The cows are gentle, and very large milkers, but the milk is poor. The Nagourie cattle are to be had in all parts of India, but I believe Delhi in former days was especially famous for them. They are very tall and narrow, with long, handsomely-curved horns; the head is thin, long and narrow, and they have a round bump standing out between their horns. They have high, narrow humps, narrow quarters, long thin tails with a very thick and long bunch of black silky hair on them. They have long hoofs and pasterns, which enable them to trot so well: it is a curious trot, as if they were keeping time to the bells generally attached to their necks, and they always go in step when well driven. Their pace is nearly as fast as that of a good trotting horse. Their colour is always of a bluish white, and they are the most delicate cattle in India; they cannot pull a heavy weight or bear hard work. The bullock carriages are like large *ekkas,* and very light, as they have two wheels only, and the pole is tilted up to prevent any weight falling on the backs of the cattle. They do not carry much flesh at any time. When intended for draught, they are allowed the whole of their mother's milk, and consequently grow very big, measuring from fifteen to sixteen hands.

English cows are not a success in the plains, but can be kept in the hills, or in mild climates like that of the Dhoon. Some years ago English bulls were brought to Bengal to improve the cattle, which on the whole are very poor in that part. At first there was a great demand for the services of these bulls, but after some years' experience it was found that the half-bred cattle were too delicate-the bullocks could not stand the sun, and the cows died when neglected in the usual Indian way; and, although they milked better than the country-bred cows, often they did not calve more than two or three times, and were sickly during the hot season. When kept in comfort by English people they were a success, but I have known natives in Bengal refuse to buy cattle when they saw the straight backs and low humps, as they recognized an English breed in them. In the hills I believe they do as well as at home, but this I have from hearsay--I have no personal experience of them; but I know the trouble and care required to keep an English cow in the plains, as she suffers so much from the climate. If you have them, they are best stall-fed; they should have nice open stalls like horses, and their grass should be cut for them; they should only be allowed to graze for an hour or two during the cool of the morning and evening. Kept in this way they are nice to have, but, what with the original price paid for them and the cost of their keep, I cannot advise practical people to purchase them. In the event of their dying or ceasing to calve, they are a great loss, as their price varies from one hundred to one hundred and fifty rupees; for this sum three good Indian cows can be purchased, and the milk of these collectively will be more than that of one English cow; and with three cows and a little management, you can have milk all the year round.

The best cows I have met within my experience are those which come from Googaira in the Mooltan District. They combine the good qualities of the Hansi cows without the great size and delicacy of the latter. Those I have met with have been middlesized, very handsome, broad and well-knit, and dark-coloured, and some of the best had black points. They were hardy and healthy, and beautiful milkers, They have lower withers than most Indian breeds, and their horns are not long.

Cows of no particular breed are most generally met within India; they are usually crosses between the common beef *gaina* and one of the good breeds, though their connection with the small country cattle bas been almost obliterated by care, feeding and breeding. Though smaller than cattle of the pure well-known breeds, they are much larger than the very small country beasts. They have no particular characteristics of their own, as they vary in shape, size, colour, and milking powers, according to the breed which predominates in their blood. They are hardy and healthy, give from four to six *seers* of milk daily, and average in price thirty to forty rupees. They generally have points common to the district in which they were born, and time and study might enable an expert to classify them, but in a short essay it is impossible to do more than given then a general notice.

COWS IN THE HILLS

In some hill-stations it is very difficult to keep cows, as food is so expensive that one cow's grain comes, to eight annas a day, and the cow gives, perhaps, only two seers of milk. In such places a buffalo is more useful, as she will eat everything, and enjoys nettles as much as anything else. The nettle is one of the commonest of hill weeds, and in Kumaon, where *bhoosa* is very scarce, the women go about gathering nettles for their buffaloes. If you do keep cows in the hills, it is best to have either English or hill cows; the latter, considering their. small size, milk much better than well-bred cows, and, being active climbers, they move easily about the hill-sides. and find plenty of grazing; while the cows from the plains miss their *bhoosa*, which is the chief food of cattle there, and when out to graze they, cannot climb about, so do not get the best grass; and they also run a risk of breaking, their necks by falling over the *khuds* if they become bold enough to attempt scrambling about. In stations where there are breweries it is easy enough to feed cattle, as the *grains* are most nourishing and very cheap. Where neither *bhoosa* nor grains can be obtained, a good plan is to chop up grass and mix with the dinners; but it is a poor substitute.

In the Hills there is a jungle grass called by the natives *Biskupra*; it makes the cattle very ill; they show great distress, refuse food and run at the nose. Two or three chattacks of *Dhunnia* (coriander seed), pounded and mixed with a couple of ounces of salad oil, has a cooling effect, and soon relieves them.

In Cashmere the grazing is splendid, and all cattle do well and milk and butter is cheap and good. In many hill-stations cows, and particularly buffaloes, will not get into calf, but become barren, the climate being unsuitable to them, so it is well to send dry cattle to the plains.

GAINEE CATTLE

This is a small breed, becoming rather scarce in India; they are not much bigger than goats, and if well fed, become enormously fat. The true breed is now preserved by native gentlemen only, the animals being kept as fancy pets; they are too small for work, and even for beef; they are pretty and look like miniature cattle: there is as much difference in size between a *Gaina* bull and a *Nagouree*, as there is between a toy terrier and a Newfoundland.

Gaina has become the term for the common country cattle, which are small, and not good milkers, but are prolific, calving once a year. They are useful for killing, and are good eating, as they are small and fatten well, being hardy, and gross feeders; and as they are cheap, it costs little to lay in a stock of them.

A *Gainee* Club is a most useful and profitable institution, and one to which I wonder domestic ladies do not oftener turn their attention, as it is not a difficult nor a disagreeable thing to manage, far less so than a sheep club, which so many undertake. Of course, at first, there are difficulties in purchasing as it takes time and experience to learn what prices should be given.

FATTENING GAINA

Gaina cows should be bought when they are three or four years old, and have from two to four new teeth; they should first be given a dose of medicine, and be tied up, grain being given to them in small quantities only, morning and evening, until they become accustomed to good food. A line of pegs should be placed out of doors under trees, and another line inside a shed for hot weather. The cattle should be picketed to these pegs with very short ropes. The best plan is to fasten them by the fore leg, as this hampers them very much. They should be allowed no exercise, which would make them hard and muscular, while complete rest makes the fat grow in layers between the grains of flesh, and produces the "marbled beef" which is so delicious. The food should consist of dry grain and dry *bhoosa*, and salt should be given once a day. They should be allowed to drink as much as they can.

As the cattle are tied up, it is easy for a lady to see them fed twice a day, and also to see that they are firmly picketed; they require little more to produce the finest beef. A *gaina* cuts up into ten chief pieces besides hump and tongue. And the skin is worth from five to six rupees. Brains, liver, heart, inside and trimmings are all worth something, and find purchasers in the bazaar butchers. If the hoofs are collected and boiled down, they produce capital soup for dogs, and an oil rises to the top, which, if taken off and strained through charcoal in the sun (see *Hoof oil*) makes a very useful preservative of leather.

The greatest drawback to a *Gaina* Club is the prejudice against it which Hindoos have. I myself have never had the courage to establish one; two or three times I have fattened the beasts, but Hindoos have always succeeded in persuading me to let them purchase the intended victims, and have paid their full price, as well as the value of the grain consumed–and were happy in the thought that they had done a good action in saving the life of a sacred animal.

An under-bred male calf also kills well if the father was of an indifferent breed. When its mother runs dry, and it is wanted no more for her to look at during milking time, give it all the skimmed milk which can be spared, and a raw egg every morning for a month or six weeks before killing, or in its place half-an-ounce of sweet oil. A calf killed at this time will not be prize-meat, but delicious eating. Bull-calves generally take after their mothers, and cow-calves after their fathers, so if a cow-calf has an ill-bred father it had better be turned into beef, as it will probably never turn out well as a milker.

HOOF-OIL

To make hoof-oil, boil the hoofs and knuckles of oxen in water till an oil rises to the surface, which skim off with care. In winter, or very cold weather, the best plan is to take it off when cold. Skim it off, boil it a little by itself; not enough to burn it black, but only to clear it. Fill a tin funnel with small pieces of charcoal (bone charcoal is best, but common charcoal will do if already used in a; *ghurra* filter); place the funnel in the mouth of a clear bottle, put them in the sun, and pour the oil in by degrees, letting it run through till clear. Hoof. oil can be bought in the bazaar, and is called *paya ka tel*, but it is always dirty. it keeps for years, and is very useful for preserving boots and shoes. It makes the leather soft and strong, and does not rot the sewing. In the dry weather it keeps the leather from cracking, and makes it water-tight in damp weather, and keeps it from being injured by, mildew. Care must be taken not to let the soles get soddened with the oil, as it softens them.

A COW DRINKING HER OWN MILK

If a cow takes to drinking her own milk, which is not an uncommon custom among some of the very best cows; or if bull-calves, for want of salt, acquire a trick of licking their own water, cradles should be put on the animals' necks. These are made of pieces of bamboo threaded together. The calf should also have plenty of salt rubbed on his tongue. The cow probably resorts to this trick to relieve the distress of a distended udder. Some calves will suck goats if they get a chance.

"GOOR," OR UNREFINED SUGAR

Natives prescribe *goor* for cows in milk, but it should not be given, as it lessens the milk, and makes the cow ready to calve long before she otherwise would. It should only be given to a cow that is wanted to calve again.

MANGE

The best cure is to rub the affected part with mustard-oil and pounded sulphur. Do not mistake prickly heat for mange.

INFLUENZA

This serious complaint is very infectious, and the affected cow or calf should at once be separated from the others. With care a cow may be saved, but a calf rarely escapes. If it does not die in the first feverish stage of the disease, it succumbs to the subsequent complications, such as sore-throat, cough, stoppage of urine, diarrhoea, constipation and sores.

The following are the symptoms: Food is refused, coat rough, nose dry, ears cold, running at the nose and eyes, general debility.

Treatment

Place the cow in a warm house; all food should be given warm; in the case of a milk- cow or buffalo put to the throat a bottle of warm beer, and hot drinks given with a bottle. For three days no medicine should be given. On the fourth day give, morning and evening, chiretta steeped in rum.

In the case of a calf, give a wine-glassful of rum in warm milk, morning and evening, from the commencement of the complaint, as the strength must be sustained. A careful observation of the patient must be maintained and the treatment varied according to the change in the symptoms. Should there be a stoppage of urine, hot tea and saltpetre should be administered. If the throat swells, hot poultices should be applied.

No person suffering from influenza should be allowed near milk-kine or young calves as animals take the disease from human beings.

Chiretta, being a strong tonic is the best remedy. It should be given with beer, which helps to sustain the animal and to keep up the flow of milk.

CREAM CHEESE

Milk, or skimmed milk, which has turned sour and congealed, or separated into curds and whey without boiling, makes beautiful cream cheese. Turn it, whey and all into a cloth to drain: hang it up, and when all the whey has drained off, leaving the curd collected into a cake, salt the latter, and warp it in a clean white napkin, and press it under a heavy weight. In about half an hour the towel will be quite moist with the remaining whey, and the curd will be left, a firm, white, sweet cream cheese ready for use.

DEVONSHIRE JUNKET

Take a pint of milk; warm it to milk warmth, and stir in a little sugar and a dessert spoonful of essence of rennet. Then leave it to set. As soon as it has set, pour over it a cup of thick clotted cream, and flavour to taste.

SOUR JUNKET

Take milk which is just on the turn, and put it into a china pudding dish, and let it stand, unless it has already set (or become *tire*); rasp some nutmeg over it, and fresh lemon-peel and a few lumps of white sugar, and add a wine-glassful of sherry or rum. Place a few spoonful of marmalade to omament the top. It is both pretty and delicious. Rum is preferable to Sherry.

ANOTHER RECIPE

If your milk is fresh, and will not turn, melt a piece of butter as large as a walnut, and stir it into the warm, milk, and then cover it and put it away to set.

VEGETABLE RENNET

The dried leaves of the flower of the thistle artichoke coagulate milk, and are the only rennet used in the South of France. The blue flower of the artichoke, fresh or dried, turns milk into excellent curd for cheese or other purposes. A tola weight of the fresh flowers soaked in two table spoonsful of hot water and strained, is enough to turn a pint of milk 2/3 of a tolah of *dried* flowers, soaked in a little hot water, and a teaspoonful of salt turn 2 quarts of fresh buffalo's milk into a rich curd.

1/3 tolah = 1 drachm.

WHEY FOR SICK CHILDREN

If you require to make whey for sick children, put a wine-glassful of Sherry in a pint of new milk, and warm it gently, and when it has separated, strain it through a piece of muslin.

TO MAKE CHEESE

How to Make Rennet

Get from the butcher half a dozen stomachs of very young calves, who have neither been weaned nor begun to eat grass. They are called *choostas*. Take out any curdled milk which may be in them, and fill them with pounded salt, and rub some salt on them outside. Put them into a wide-mouthed jar, and cork tightly. They will keep good in this state for a year, but will be ready for use in a month. When cheese is to be made, take half a *choosta*, cut it up into thin strips, and put in a jar with a *kaghzi* lime cut up and half a teaspoonful of cloves pounded coarsely; over these pour a pint of tepid water, and cork it; give the jar a shake every day for a week. It is now ready for use.

How to Make Cheese

Take two spoonfuls of rennet, which will coagulate twenty *seers* of milk. Heat the milk to 85° in summer, 1000 in winter, and strain it into a wooden pail; strain the rennet into it, and in cold weather cover over with a blanket near the fire. If the rennet is good, the milk will have set or coagulated in a few hours. Cut it across down to the very bottom of the pail with a wooden knife; cover up again for another hour, by which time the curd will have settled down to the bottom of the pail, and the whey will be at the top. Strain both curd and whey through a coarse, but thin cloth. Tie up the cloth, hang it up, and let it drain all night. The next morning put the lump of curd in a dish or large milk-pan, and divide it into large pieces, adding four ounces of salt to every twenty *seers* of milk originally set. This quantity makes a good-sized cheese. Moulds or small barrels should be made of cask staves, eight inches deep and six inches in diameter, and bound with iron hoops. In one of these place about half-a-yard square of coarse but thin cloth, and put in the curd gradually, pressing it down gently with the hand until it is full. Fold the ends of the cloth over the curd, and tuck in well all round, the edges inside. A number of circular wooden discs should be made to fit inside the barrel, some thin, some thicker. Place one on the cheese, and press it down gently; then another, and so on. This should go on for a couple of days, by which time the cheese will have hardened, and will bear taking out of the mould, cloth and all; the barrel should be gently hammered all round in order to loosen the cheese. Lay the cheese on a shelf, and turn it every day for a month, by which time it will be ready for use; but the older it gets, the nicer it will be.

The milk, while being heated, should be coloured with saffron in the following manner: steep twenty seeds over-night in a wineglass or cup full of cold water; in the morning rub the seeds, and strain into the milk while being heated. The price of saffron is from six to eight rupees a *seer*, but a *chittack* goes a long way.

The above recipe was given me by a lady famous for her delicious cheese.

Cheese can be made in very cold weather only.

TO MAKE "*GHEE*", OR CLARIFIED BUTTER

Clarified butter is known in England as "French butter," and is to be had in pretty white bottles. Let; it be made before you in a back *verandah*, or some such convenient spot. For some days collect all the spare butter, and when you have enough to boil down, put it in a small round *Deghchee* on a clear charcoal fire, *not too hot*, and let it boil *gently*. At first it will froth up like whipped cream; then by degrees it will curdle, and small white opaque particles will be observed boiling about inside; after a time these particles; begin to adhere to the bottom, leaving the *ghee* quite clear; when it is nearly ready, it will leave off frothing, and will simmer quietly, rising in round glassy bubbles; now stir with a spoon, so that the froth at the top may go down, but do not disturb what has caked at the bottom, which, after a time, will look like a poached egg, while all above will look like salad oil, clear and bright. It is now done; take it off the fire, and pour it gently into an empty tart-fruit bottle, and, when cool, cork down and cover with bladder which has been soaked in a little blue-stone and water; otherwise the ants will eat off the bladder cover. As the *ghee* cools, it will become feathery, and look like honey. If carefully made, it will keep good for years. The great secret in making it is patience; it must not be boiled on too hot a fire, or it will burn, turn brown, and smell very nasty. Home-made *ghee* goes twice as far as butter in cooking, for it is condensed by the previous boiling; and if it is well made, it tastes, when used for cooking, just as nice as butter, and is very wholesome. Bazaar *ghee* is nasty adulterated stuff, and not properly clarified. If you have to buy it, you should clarify it as follows:

TO CLARIFY BAZAAR *GHEE*

Add to it a cupful of sour milk or butter-milk, which will curdle up in the boiling *ghee*, collect all impurities, and settle down at the bottom, leaving the *ghee* clear.

REMARKS ON *GHEE*

If *ghee* has been kept any time, and there is a doubt about its sweetness, turn it out of its bottles, warming them first, and reboil it, at the same time putting in a cupful of milk, a teaspoonful of salt, and a few cloves and cardamoms. Reboil in the same manner as described already, and it will become as fresh and sweet and clear as newly-made *ghee*. If anything it will be milder, possessing scarcely any flavour, and consequently excellent for cooking. Mohammedan cooks are very fond of using bad *ghee*, or beef suet; one is offensive, the other is unpalatable as it cools.

When good *ghee* is not procurable, mutton suet rendered down is excellent, as it is wholesome, and yet does not clog when cold, like beef suet *Mugs*, and Hindoo cooks generally, use it in preference to any other grease.

Hill people have a harmful practice of collecting their *ghee* in large copper *ghurras*, and, when they have sufficient, bringing it to the bunniahs for sale. It appears to be good, but as soon as it is put on the fire it assumes a greenish colour. Such *ghee* must be very harmful, and should be rejected.

A GOOD RECIPE FOR SALTING BEEF

2 ounces salt.

4 ounces brown sugar.

A little vinegar or a lemon.

Rub the above into a round of beef. On the 4th day tie it up in a cloth, and boil very gently until *quite* tender.

The beef will be delicious.

SWEET JUNKET FOR CHILDREN

Take a quart of skimmed milk and warm it, then stir in enough sugar to make it sweet; then mix a table spoonful of essence of Rennet (Clark's), and cover and put aside to set.

The same made with a little lemon peel grated, or vanilla to flavor it, and clotted cream poured over it when set, is a delicious pudding for grown people).

THE BEST SCONE FOR AFTERNOON TEA

½ lb. Delhi flour (4 chattacks)

1 chattack of butter.

1 teaspoonful of baking powder.

The yolk of 2 eggs.

½ teaspoonful of sugar.

½ teaspoonful of salt

Beat 15 minutes, adding milk as required to make a stiff paste. Five minutes for baking.

Part II

The Management of Fowls in India

Chapter 1

Lewes Wright, in his beautifully-illustrated work on poultry, commences his book by saying: "Success in poultry-keeping must obviously depend very much upon the character of the premises the fowls are to inhabit; they cannot, any more than human beings, retain health and condition in unhealthy dwellings.'

Accommodation

The first thing is to get a suitable house for poultry, as success depends greatly on this. It is, of course, best to build one specially adapted for housing poultry; but as residence in India is generally migratory, it is not always convenient to do this, and then a spare out-house will have to be occupied.

In the Plains

Having selected your house, let it be thoroughly cleaned and *leeped*. If very dirty and old, it should be re-plastered, the old plaster being removed first. If the house has been previously occupied by poultry, burn all the old roosts, boxes and other woodwork. Let the doors and the lintels of doors and windows be thoroughly washed, and then painted over with kerosine oil, a large brush being used. If the floor is of mud, have it dug up and relaid; if of cement, it should be flushed with clean water, and then whitewash or quick-lime should be spread over the whole room, care being taken to run it into every crack, corner and cranny, so that all vermin-eggs may be quite, destroyed.

Also carefully burn and destroy all old *jaffery* work round about the house, as the germs of disease often hang about such work for months, and, in the spring and rains, reproduce sickness among the new and healthy poultry.

New *jaffery*, or wire-netting, should be put round outside the door in order that there may be an enclosure in which to shut up the fowls during the heat of the day, and during laying hours.

A new-house is, of course, preferable to an old hut cleaned up, and if you have decided on the expense of making one, build it 8 to 10 feet square, according to the number of fowls you intend to keep; put in front a door 3ft. × 6ft. Panel for two feet up, and, above this, place an open panel covered with strong wire to admit air. In the back wall place one or two windows fitted either with iron bars or strong wire netting. In the hottest parts of India, it is desirable to have windows in the side walls also, so that a free current of air may pass through the house. By nature fowls roost at night on trees. where they enjoy the cool and fresh night breezes. The windows should be very large, as small openings produce draughts, which are harmful to poultry; the more open the fowl-house is the better. Strong roosts should be put high up across from wall to wall, not exactly opposite door and windows, but a little above or below the latter, so as to avoid draught. The floor should be *katcha* to admit if its being dug up from time to time, and relaid with fresh earth. After plastering with cow-dung, spread over the floor a layer, a couple of inches thick, of loose earth. This soft dust is the best and cleanest material for keeping down lice, for the hens rub themselves in it; it is easily brushed up, and replaced when necessary; it also keeps the house from smelling badly owing to the droppings of the fowls, and the feet of the poultry clean.

Boxes, made like dog-kennels, should be placed in the corners of the house. The openings should not be large, as hens love privacy; place in each box a nest-egg made of wood, and painted white with Aspinall, or white marble eggs from Chunar or Benares; these are the best, as they resemble real eggs, and last for ever if not stolen. The natives make these stone eggs for their *poojahs*, and they can be obtained in most Hindoo towns. In each box place fresh grass with an egg.

In England, nest-eggs are placed in the nest to entice the hen to lay in the desired spot: when you remove the real egg, the hen does not miss it, but goes back daily to lay in the accustomed place.

There should be a small trap-door low down in the panel of the front door of the fowl-house, to admit the hens while the house is locked up. Take care that this trap-door is not too large, or the sweeper's children may creep in and help themselves to the eggs. A few nails in the upper side of the opening, with their points sticking out, prevent petty larceny, and do not hurt the fowls!

The Fowl-house Yard

This should be of a convenient size, and enclosed by *jaffery*. It is required to keep the fowls penned up during the heat of the day, and when the hens are laying. Unless they are shut up between the hours of 10 and 2, they will be decoyed away by the servants, to lay in the stable or in a servant's house. During the laying hours they should not be disturbed. Country hens generally lay between 10 and noon, English hens from noon to 2 o'clock, and half-breeds at any time between 10 and 2.

Fowls should be let out as early as possible in the morning, and at 10 o'clock fed inside the yard. This arrangement makes it easy to shut them up.

In the cold weather, particularly in Bengal, where the dew is heavy and does not vanish till the sun is well up, the fowls should not be let out till 8 A.M., or they will become ill from damp and cold feet. Damp and cold are always, bad for fowls, and during the rains the more they are penned up, the better. Young chickens should *never* be allowed to run about loose on a wet day. The confinement is irksome to them, but is necessary for their health. At this season of the year, too, fowls seem to be an easier prey to the jackal; the latter seems to know that the poor birds are at a disadvantage from the weight of their wet feathers, and heavy and awkward in consequence.

The *jaffery* should be about six feet high. in places where it is not easily obtained, wire netting, will do. If you do not put wood at the top of the netting, you need not have it higher than three feet, as the fowls, being unable to tell the height of it, will not attempt to get over.

The fowl-house should be built near the stable, and as far as possible from the garden. Fowls love scratching in stable-litter, and will seldom leave the dung-hill to injure the garden.

It should not be attached to the stable, as in that case the insects from the fowls worry the horses, but the former scratching about the stable and stable-yard can do no harm.

Ducks should not be kept with fowls, as they dirty the drinking water, and paddle about with wet feet and feathers, until they make the place unfit for fowls to live in.

An iron pan for drinking water should be provided. Raise it off the ground on stones, so that the fowls can drink without getting their feet in to the pan. I recommend iron for the vessel because it rusts the water, which is thus made strengthening. *Pucka* mortar should be in the corner of the yard for the fowls to pick at.

Fowl-house in the Hills

It is simpler to build a fowl-house in the hills than in the plains, for it need not be so high, nor need it have so many large windows, and, if the door has wire netting, there is no necessity for more than one window, which should be on a side wall, and not opposite the door, for fear of draught. The window should be glazed; and made to open and shut, and have wire netting outside, so that the temperature may be regulated as required: in the winter, boards should be nailed over the wire netting of the door, as, otherwise, the cold inside the house is too great. The yard-should have an iron roof, so that the fowls may be penned up throughout the rains. They should be allowed out only on a fine day. All the other arrangements should be as recommended for the plains.

It is best to remove the eggs every afternoon, for, if they are left all night in the boxes, there is a chance of rats carrying them off; this is often the case in the hills. I have often found empty egg-shells in the rats holes and store-dens, together with other articles stolen by these destructive and hardworking little animals. Maize should be kept in closed boxes, as rats will carry off as much as a *seer* in one night. "Rough on Rats" is the best cure for these troublesome creatures; but as it also kills fowls, great care must be taken to open up the rats, holes, and put the bread, with butter and poison spread on it, into them, and then close up the holes.

For the treatment of poisoned fowls see *Poison*.

Chapter 2

Choosing Fowls

After preparing your house and getting everything ready, buy a number of common country hens, choosing those with short legs and heavy bodies, and as many black, brown and speckled as possible, rejecting *snow-white* ones (Albino), as they are delicate, to rear and tasteless for the table. Also avoid the game breeds, as they are bad layers, sometimes giving only from three to five eggs at a time, and they will destroy the harmony of the yard, for they fight from morning till night, the hens being as quarrelsome as the cocks. They are good eating while young, but not ornamental when trussed, as they are all legs and necks, much the same as the large Chittagong poultry, who have legs quite out of proportion to their bodies.

The Negro

One of the best fowls in India is the "Negro," or small, black skinned fowl; it is seldom used by *Khansamas* for anything but cutlets and curries, as, on account of its black skin, it does not look well on the table but it is by far the best flavoured fowl in India; when well fed and kept, it has much the favour of a pheasant. It is small and a good layer; the hens sit well, being strong, with short and slight legs, which do not trample the young chicks. They are not always black, their feathers being often of a dark grey, but bill, legs and skin are of a blackish hue. Even the bones are sometimes blackish, and the eggs, when hard boiled, are of a blackish blue round the yolk. All spotted poultry are both good and pretty.

Mixed Breeds for Domestic Use and Economy

English poultry has, for so many years past, been imported to India, and the breeds spread among the original country fowls, that it is common now to find among bazaar fowls some that have top-knots, feathered legs or whiskers, or that bear other traces of English blood. Such chickens are the very best to buy, as they grow into fine fowls with the good qualities of both races. It may be asked: "What good quality is inherited from the country *moorghi* ? "This much-abused bird is one that is most essential to success in poultry-keeping–it has such vitality that it does not die readily, even under the treatment it receives from natives, so it gives stamina to the over-bred, delicate English bird; it does not suffer from the climate as birds of purer breed do, and consequently fowls of mixed breed are the best for domestic purposes. A few English or half-bred cocks among country hens will produce large eggs and strong chickens, all good for the table. One cock should have five hens, thus making up the half-dozen, or two cocks in twelve birds. A larger proportion of cocks produces disturbance in the yard and harassment to the hens–a smaller proportion is not enough–the, eggs will become small, and decrease in number. Cocks are delicate than hens, and soon die, particularly of liver, so it is well to have a few young birds in the yard, growing up with the older ones, to fill vacancies as they occur. If English hens are mated-with country cocks, the eggs become small and of a white colour; so I advise the provision of a few English cocks, if procurable and a good number of nice country hens. Even if all the birds are English, it is well to admit a strong staff of country-*hens*, who prove useful in sitting on the eggs of their English sisters, who are useless for this purpose: if they sit out the time, they will trample and crush more than half the chickens; they are far too awkward and clumsy for mothers.

Quarantine

One precaution is very necessary. On the other side of the compound set apart a house for newly-bought poultry, which should be kept separate until it is quite certain that they are free from vermin and contagious disease. The admission of one sickly fowl or chicken may prove disastrous to the whole stock, so I cannot too strongly impress on my readers the necessity for this precaution. When natives discover that a contagious disease has broken out among their poultry, they promptly dispose of the whole stock, which is bought up by *khansamas*, and converted into food for their masters before the contagion can show itself; but these birds admitted into a clean healthy poultry-yard will infect and ruin it.

Care should also be taken that no bazaar poulterer, whose clothes and baskets may carry infection. be allowed near the yard.–(See *Roup*).

English Fowls in the Plains

These beautiful birds do well in India, but require great care during the hot weather. With plenty of fresh air *at night* and protection from the sun during the day, they thrive. They cannot bear close pens at night, as they have heavy hot plumage which distresses them. As they are delicate from over-breeding, they are subject to liver complaint. Damp floors will produce enlarged livers, from which they die suddenly.–(see *Liver*).

The sun also affects them, and they will avoid it if they can, and will sit in shady places, but perhaps only after they have been distressed by it, so it is best to provide a place where the sun cannot reach them, and to let them have a run in the afternoon, when the sun has lost its power. They require plenty of fresh air, so the place provided for them should be open to the breeze.

English Fowls in Hills

In the hills, those who are fortunate enough to possess English fowls seldom find them anything but a success; it is, consequently, worth the while of those residing for any time there to go to considerable expense and trouble to procure the best breeds. As the climate suits the birds, and they are not subject to vermin to the same extent as in the plains, they give comparatively little trouble. Strange to say, the common house-bug, which infests servants houses and all places in which the hill people, who are dirty folk, live, is a cause of much annoyance to the English bird. It gets into the wood-work of the fowl-house, and should be guarded against.–(see *Bugs*).

English fowls lay splendidly, and only stop at moulting time when they require a little care and protection from chills.–(see *Moulting*).

If well led on soup-meat and hot meal of any kind, they give eggs throughout the winter. Rice should be avoided, as it causes diarrhoea.

Preservation of Eggs

Eggs can be kept quite fresh and good by smearing them over with fresh, sweet butter, and placing them in a box in a dry, cold place. When wanted for use, they should be washed with warm water, after which they look and taste like new-laid eggs. As eggs take a taste from any smell near them, be careful to keep them where no strong smell exists: for instance, if eggs, particularly new-laid ones, are placed near onions, they will receive a decided oniony flavour. This should be remembered in storing them, and also that the butter with which they are smeared is fresh. *Ghee* preserves them as well as butter, but gives an un-pleasant flavour. So does keeping them in straw. If you wrap them up before you put them into a box, use silver or tissue paper.

Price of English Fowls and Eggs

As the eggs of English fowls sell at Simla for Re, 1-2-0 per dozen, a little profit, as well as pleasure, can be obtained from the farm-yard. Young chickens fetch one rupee each, and full-grown fowls about five rupees the pair-three rupees for the cock, and two for the hen. The price of thorough-breds is ten rupees for the pair.

Brood Hens

When English hens become brood, or, as natives call it, *crook*, they should at once be removed from the hen-house, as, if left, they will destroy the eggs of the other hens; they also fight with the hens coming in to lay. They should be shut up in a strong pen in the open air, and not allowed to go near the laying-box for a week; when they leave off clucking they may be let loose. If allowed to sit when in this state, they often die, their state of health and the heat of their plumage being too much for them. When set on eggs, they often die in the nest from distress.

Jackals and Foxes

In the hills the fowl's worst enemy is the fox; on this account the poultry should be penned up early in the afternoon. During the rains, both in the hills and plains, the jackals also prey on the fowls. When the winter is coming on in the hills, both jackals and foxes must be guarded against. The fowls are never taken in the morning, but always after 2 in the afternoon.

Chapter 3

The Country Fowl or Murghi

The despised and unpalatable, but long-suffering *murghi* requires a chapter to itself, and a few words in its behalf. It is only known to us in India as being a tough and tasteless article of food, which poor Anglo-Indians are forced to eat, dished up, without regard to the most elementary culinary rules, by every *khidmutgar* and *khansama* from Cape Comorin to Chitral. I once heard a witty remark from a man who was invalided. When asked what his complaint might be, he said he was suffering from *murghies*, and that they had finally succeeded in sending him home. This shows how a long and systematic course of mismanagement may bring universal contempt on a good thing. The poor *murghi* may well exclaim: "Give a dog a bad name, and hang him."

In reality the Indian fowl is a delicious bird with a delicate flavour, but it must be properly fed, kept and *killed*, as I hope to explain to my readers. The birds are small-boned, full-breasted, possess finely-grained flesh, and are capable of unrivalled fatness, but the long-suffering creature, as we know it, only continues to exist in virtue of its hardy constitution. Any other animal, subjected to similar treatment, would have died out ages ago. It is only since English people have come to India, that it has obtained a market value. Maohmedans certainly eat fowls, but they prefer goat and beef. Hindoos look on the bird as unclean, and only keep them to offer up in sacrifice to devils and evil diseases, such as small-pox, to propitiate *Kali mai*, the evil genius of cruelty. In former times, two or three fowls were tied to the bed of a small-pox patient, and when he recovered, the fowls were let loose in the jungle, or drowned in the holy river, the Ganges. If the patient died, the fowls were slain with cruelty. Maohmedans value the game-fowl, and breed it carefully, as they gamble largely on cock-fights.

The *murghi* hatched and reared for the market is usually kept in the smallest of hovels, often built in a sunny spot, and of an oven-shape, with no opening for air except the door, which is jammed up with bricks to prevent snakes and cats from taking the poor creatures. Such a building is generally swarming with vermin, which prey all night on the birds, impoverishing the blood and disturbing the rest of their victims. After a night of suffocation and horror, the poor creatures are released in the morning, and horror off to any water there may be about, and then drink and drink as if their lives depended on slacking their raging thirst. When they have recovered a little from the sufferings of the night, they wander about in search of food, as few natives feed their poultry, who can generally pick up enough to keep body and life together, which is all their master cares about. When they are old enough to fetch a price, they are taken to the bazaar, and sold to a cook, who carries them home, hanging by their legs. When required for the table, he catches one, takes a knife, turns his face in the direction in which he supposes Mecca lies, says "Bismillahe Alla O Akbar, "gashes its throat, and flings in out towards the sacred Mecca. The poor creature is then considered fit for food. It kicks and struggles and flutters and bleeds until it dies, as much from suffocation as loss of blood, as the cut is neither deep nor wide.

Dressing the *Murghi*

When the poor bird is taken up, it is rigid and stiff. Every muscle seems to have become wire. The other day I was reading a book on health, written by an eminent doctor, and was pleased to learn on such good authority that improperly-killed meat was most indigestible, as when the animal had suffered much, the flesh became hard and attenuated, so my witty friend may have been right when he declared that he had been sent home with a ruined digestion by the Indian *murghi*.

As soon as the fowl is dead, the cook plunges it into boiling water for a few minutes, which effectually deprives it of any juiciness and fat which it might have possessed, for the fat of a fowl, like that of a pig, grows close to the skin. He then rubs it down violently, by which means he clears off the feathers, skin and all. Then he cooks and dishes up, not unskilfully; the native chef originally learnt the art of cooking from the French, and as all natives pass on their trades from father to son, we still enjoy the benefits of French cookery but somewhat disguised by the brutal manner of killing and short cuts for plucking (which is a tedious process, and has to be performed with care) which the native has introduced to suit his own feelings and convenience. Chicken cutlets and grilled fowl (Spatch-cock) are exellent dishes if made of properly fed, killed, plucked and hung poultry. (See Receipts in "Chapter of Chapters.")

In this little work I cannot enter into a description of Fancy Poultry, for full particulars regarding which I refer my readers to a beautifully-illustrated work: *The Illustrated Book of Poultry*, by *Lewis Wright*.

There they will learn all about the different handsome breeds which are to be had in England; their names, prices and histories. I can mention only a few breeds easily obtainable in India, and whether they are or are not useful for domestic use.

Chapter 4

Laying

Some country hens lay, but most of them every other day. They give from six to twenty-five eggs at a time English hens give from twenty-five to fifty, half-breds from ten to fifty, depending on the crossing. If the hens are English and the cocks country, the eggs will lose size and richness of colour, while if, on the contrary; the hens are country and the cocks half-bred, the eggs will be large and of a rich colour.

Large English cocks and small country hens should not be paired; their sizes should be equalised as nearly as possible; otherwise the eggs are apt to be unfertile.

The faces of laying hens are red and hot, and the birds run with difficulty, and should not even, as it makes them lay soft eggs an hour or two before the time (see *Soft Eggs*). If the cocks do not crow, the hens are not laying. As soon as a hen has laid, and rested on the egg for a short time, she runs away, and foolishly gives the alarm to the whole yard, so a listener can at once tell if an egg has been laid. I believe a hen does this to call away the public attention from her nest and egg. In a similar way a bird, who has built her nest in a hedge, will flutter along in front of a passer-by, as if she was hurt, to distract his attention from the little ones in the nest. When hens have done laying, their breasts become bold and featherless, and they commence making a sound of "cluck, cluck, cluck." Hence the appellation of *cluck moorghi* for a brood hen. Hens often lay one or two more eggs after becoming broody, which shows that they begin clucking a day or so before they are ready to sit.

Chapter 5

Food

Indian corn is the best and cheapest food for fowls; they lay well on it, and grow very fat. *Bajra* and *chota jawar* are also suitable grains.

Laying hens should have a morning meal of hot bran and *urdawa*, either boiled, or after boiling water has been poured over it: it should be stirred until it cakes and becomes porridge; in very cold weather chopped onions should be added to this hot mash.

In the evening, before shutting up, a feed of Indian corn should be given.

Green meat is very necessary in places destitute of grass and herbage. Vegetable-tops and trimming, chopped very fine, should be put into the hot mash. Table-leavings are invaluable for poultry, but seldom given, as the sweepers keep them for themselves.

Poultry brought up to the hills should be fed for the first few days on a mixture of *atta*, ginger, black pepper, mustard-seed-oil and chopped onions; this prevents their catching cold from the change in temperature, and is particularly necessary for chickens arriving in the hills in the very early spring and in the cold autumn. English fowls should, on no account, be fed on barley, as it gives them a disease, in the crop, which kills them.

Chapter 6

Eggs for Setting

In choosing eggs for setting, select those of oldish hens, as their chickens are strong. The eggs of pullets hatch, but the chickens die while getting their feathers. In selecting eggs for setting it is essential to remember that they should have come from old birds. In every setting half the eggs should be large and oval, half small and round, the former will produce cockerels, the latter pullets. Eggs for setting should be removed from the fowl-house as soon as the hen leaves the nest, to prevent their being kicked about and bruised by other hens. They should be taken from the nest by a hand which is warm and dry. Eggs for setting should never be touched by a moist, cold hand. They should be carried with as little shaking as possible, and they should not be allowed to strike against one another, or they may be slightly cracked, which may result in their becoming addled. If you have an egg-frame, place the eggs in it, the air-bubble uppermost and the narrow end under. A box-full of dry sand or saw-dust is equally good. Place the eggs in it a little apart, and turn them daily, to prevent the yolk from sinking to one side.

As soon as you have collected a sufficient number, set the eggs. They should not be more than a week old. After this time the white skin or membrane between shell and chicken becomes dry and hard, and, when hatching, the chicken is unable to free itself, and is strangled. Should this happen, the chicken must be helped out of the egg, but it cannot occur if the eggs are *fresh*. Should the eggs be known to be rather stale, set them in a damp place, so that steam may be engendered by the hen sitting: this will soften the skin to a certain extent. A thunderstorm will often addle incubating eggs; it is said that eggs in an incubator are not similarly affected by changes of the atmosphere.

Time for Setting in the Hills

In the hills the eggs should be set as early as possible, in February and March if practicable, so that the chicks may be strong before wet weather sets in, and the moulting season arrives, which is always a sickly time for chickens; if they have not grown to a good size when the monsoon begins, they will die off from damp; if the settings run into the hot weather, the eggs will spoil.

Time for Setting in the Plains

In the plains chickens should be out in December and January, as later they will be disturbed by insects. Chickens do best in a temperature of about 60°F.

I must not be understood to say that eggs set at other times will not be a success, but only that these two months give the best results.

Sitting Hens

If the poultry-yard has, as I have advised, both English and Country hens, the sitting becomes easy, as they all will lay about the same time, and the country birds become brood just when they are wanted to sit on the English eggs; a hen always sits best in the place in which she has laid her eggs. *Crook* hens from the bazaar are very. troublesome, and seldom give satisfaction.

An old shawl or blanket should be spread in the box in which the hen is to sit, as it helps to keep the eggs warm. The box should be just large enough to allow the hen to sit comfortably, and not permit of her scattering the eggs. Place a large bamboo frame (*tappa*) over it, leaving room all round for the hen to get up and walk round to stretch her legs. A pan of water and some dry maize should be placed on the ground inside the frame. She should be given no other grain, as sitting hens are prone to loose bowels which makes them soil their eggs, and produces many difficulties which are avoided by feeding on dry maize.

English hens should never be made to sit. They sometimes forsake the eggs before they are hatched; with their heavy limbs they crush the chicks which are coming out; and occasionally they die on the eggs.

Half-breeds are sometimes a success, but often not so. The small, thin-legged country hen is almost always a reliable mother. A good mother is worth keeping for sitting purposes alone, but after once bringing out successfully a sitting of chickens, she will always be troublesome afterwards, when brood, if not allowed to sit.

From seven to nine eggs should be put under each hen. It is unwise to put more, as by turns they get away from the warmth of the mother, and so the whole sitting may become addled.

Every hen should have her own box of eggs and her own *tappa*; if not imprisoned thus, some hens, will leave their own eggs to fight or disturb the other hens.

Testing Eggs

Home-laid and home-set eggs seldom require testing. Bought eggs are unreliable, and should be tested on the 7th or 8th, and again on the 14th or 15th days. For this purpose a testing lantern or an ordinary kerosene lamp may be used. The latter should be enclosed in a wooden box without lid, having a hole about three inches in diatmeter cut out of one side directly opposite the flame. The egg-tester, fitted with diaphragm, should he held in the left hand between the light and the eye, so that the egg, which is held by the fore-finger and thumb of the right hand, close to the egg-tester, intercepts the light. The whole of the contents will now be clearly seen. Those eggs which are decidedly fertile will be found to have a dark patch in the centre, and a number of red specks running in various directions radially, giving the embryo the appearance of a spider with long legs. An unfertile egg has the same appearance as a new laid one.

A little experience is necessary for the reading of eggs through the shell, even with the help of a lamp. Egg-sellers are so expert that they can tell the state of an egg by holding it in the sun's rays, making a cylinder of the left hand while holding up the egg, with the right, opposite the eye.

A hard-boiled egg is easily discerned by spinning it on a table; if hard, it will readily spin; if soft-boiled it will spin, but not with the

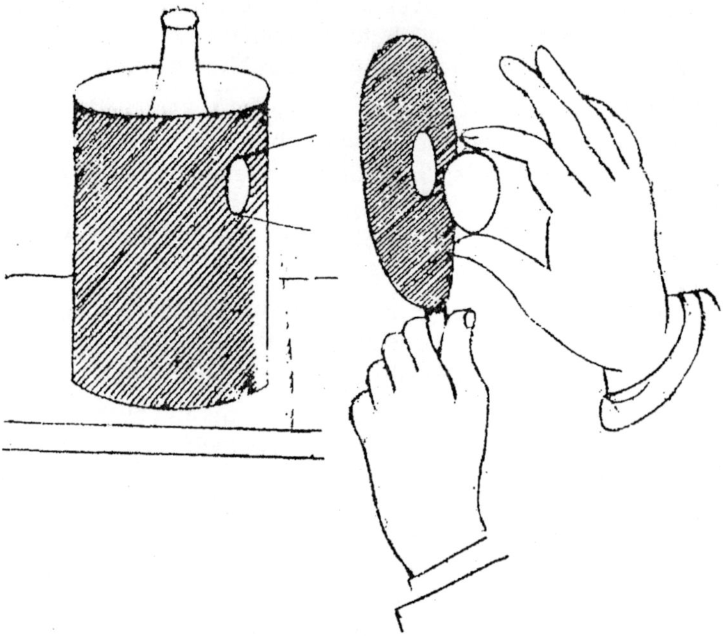

same velocity; an uncooked egg will not spin at all, as motion is checked by the fluid inside it. It may seem unnecessary to mention how a hard boiled egg may be tested, but in Kabul and all Pathan countries the natives gamble with them, shooting the eggs at one another like marbles, the one that cracks losing the game. For this reason hard boiled eggs which have failed to obtain a sale for gambling purposes, are often mixed up with unboiled eggs in the hope that they may pass undetected. Cooks, too, often mix up their eggs and are perplexed as to which are the boiled ones.

All the above information is useful to persons who have incubators, and is to be found in the little pamphlets which are sold with the machine; and it may also be useful to others, who like experimenting with hens and eggs. It is, however, safer if the stock is good, the eggs well chosen and the instructions followed, to leave the hen and eggs *alone*, as much as possible.

Chapter 7

Hatching

Hens' eggs hatch after	21 days
Pheasants hatch after	24 days
Guinea fowls hatch after	26 days
Ducks hatch after	28 days
Peafowls hatch after	28 days
Turkeys hatch after	28 days
Goose's hatch after	35 days
Ostrichs hatch after	40 days
Swans hatch after	42 days
Emu's hatch after	60 days

Climate may cause a little variation in the above-stated periods. I have set and hatched chickens in all parts or India–hills and plains–and in England, and I have found that in a hot climate eggs hatch sooner than in a cold one; there may be as much as a day or two's difference. On the twentieth day the eggs of hens generally begin to chip, and most of the chicks, if strong, will be out in 12 hours-sometimes they may be as long as 18 hours. A couple of days before the eggs chip, the hen will begin to cluck as if she were speaking to the chicks, and, when she moves, the eggs will give out a rustling sound, as if they were empty egg-shells rubbing against one another, and, if you take up one of the eggs, it will feel heavy, and the shell will look white and dead, for the chicken has left the shell. Sometimes it can be heard calling inside the shell, and its *cheep cheep* may be distinctly heard as soon as the smallest crack appears on the shell. If the chick is caught and strangled in the white skin which lines the egg-shell, a little help will relieve it; if slow in clearing itself of the egg-shell, the middle of its body may get caught, as it expands all over as soon as the air reaches it–in this case also it will require help. Should a newly-hatched chick get away from the warmth of its mother, and appear to be quite dead, vitality may be restored, even if it has become quite cold and stiff, by putting it back under its mother. On the night of the 20th day take up the hen, and hold a bright lamp over the eggs for a few minutes, and then replace the hen. This helps the chickens to come out all together, though I cannot tell why it should do so. But if you have a good mother hen on the eggs, it is best to disturb them as little as possible, for, as soon as her hopes are raised, she will naturally keep perfectly quiet and will not interfere with the hatching by rolling about, or otherwise disturbing, the eggs.

In a small work like this, it is not necessary to follow the chicken through all the wonderful changes which take place inside the shell, from the moment of incubation to the hatching of the egg; interesting as the process is, a description of it would be of little use to the practical poulterer. But it is as well to know that the absorption of the yolk takes place within the last twenty four hours, and that it supplies the chicken with sufficient nourishment for thirty-six hours after hatching, so that there is no necessity to disturb the little ones for the first two days. On the third day they will be observed to put out their downy heads, and then it is time to take them out and feed them, first noticing if the little knob on the tip of the bill, with which the shell is cracked, has fallen off; the hen will often pick this off, as it looks like a little grain, and attracts her attention. Should she not have removed it, it should be picked off with the nail.

Chapter 8

Food of Chickens

This little work is merely a handbook of useful practical hints on poultry and their management for domestic purposes, and I can guarantee success if my instructions are fully attended to, as I give no hints the truth and success of which I have not proved. As much that I advise is contrary to the advice given in other books on poultry, I fear that I lay myself open to hostile criticism, but I am so sure of my facts that I ask for nothing more than a fair trial of the treatment I prescribe, and am confident that those who do fairly adopt it, will be converted to my opinion. What I am now about to write contains the whole secret of success in rearing chickens, and is a secret which I learnt by accident. It is kept by professional poulterers, and is not to be found in books.

The digestive powers of a chicken are marvellous, and are developed during the first fortnight of its life, at which period the harder the food it gets, the more powerful its digestion becomes; the chicken resembles the captive ostrich, which dies of indigestion if not given iron nails and stones to keep its digestive powers in proper work. Hard boiled eggs, bread and milk, crumbs, and such like soft diet generally recommended are unnatural, and therefore to be avoided; chickens fed on improper food of this description become weakly, and when their feathers come, they have no strength to bear the strain on their system, and die of the various disorders which attack half-grown chickens, and destroy them wholesale, such as diarrhoea and *cheeps*. Most poulterers know this secret, but will not divulge it. I learned it from a poor man whose livelihood consisted chiefly of what he made out of his fowls. When I asked him how he made it pay, as the feeding of chickens was so expensive, and so many died, be replied curtly: "I cannot afford to kill my chickens with hard-boiled eggs and such trash." This gave me the hint I wanted, and, on experimenting further, I discovered that hard dry grain was the proper food. Maize, or *mukkee*, as the natives call it, is the best food from the very first, and I have reared scores of strong healthy chickens on nothing else. When they are very young, it should be crushed a little with a stone, so that it may break up easily. Should there be a very weakly chicken that will not pick up the food, take it up and push big bits of maize down its throat. Its digestion has not commenced to work, and requires starting, which this treatment will effect. At its next meal, it will be able to feed itself. The droppings from chickens fed like this are dry, and have no smell, and can be easily brushed away; while chickens fed on what my friend called "trash," are very dirty and objectionable. When the chickens are a few days old, they may have chopped raw meat which will take the place of the insect diet which is their natural food.

All poultry lay well and grow fat on maize–ducks and geese as well.

If chickens are to be brought up by their mothers it is well to allow them to scratch about out of doors as much as possible, as they pick up numberless things which are good for them, of which we do not know. As they grow older, a box, without top or bottom, should be made for them: the top should be covered with wire netting; put the hen and chickens inside, and have the box pushed from place to place, which will give the chickens fresh ground for scratching. Wire netting is the best protection against cats, but not against the mongoose, unless the mesh is very small.

A little iron is good for chickens when their feathers are coming; for this reason rusty nails should be placed in the drinking pan. Chickens are best reared in dry climates, and at a temperature of over 60° F.

Chickens Reared by Hand

It is best to bring up chickens by hand if the necessary time can be devoted to them. The following arrangement will be found useful, especially to those who possess an incubator as it renders unnecessary the employment of a foster-mother and other devices connected with the rearing of motherless chickens.

As soon as the chickens are hatched, and have been long enough with their mother to be quite dry and strong, and have a little warmth in themselves, take them away from the hen, wrap them up warm in a nold shawl or other woollen article, and put them into a dry, clean basket, in a warm place inside a cupboard or deal box; taking care that the box does not shut too closely. Young chickens do not require much air only enough to breathe. Two or three times a day open out the shawl, put the chicks on some dry sand in the sun; place a shallow saucer, full of water, near them, and sprinkle some crushed maize on the ground. At first the chickens will require teaching how to pick, and the best way to instruct them is to tap on the ground with the finger in imitation of the hen, who always taps with her beak when she wishes her chickens to pick up what is there. They soon learn but if slow to do so, must be fed by hand for a day or two. After a few days they become quite independent, and run about, and pick up whole grains of maize. After feeding, they should again be wrapped up and put away. Chickens reared in this way are perfectly clean, and have no smell, and it is a pleasant amusement to bring them up. A little chopped meat is very necessary, but it should not be salt meat, as nearly all salted meat contains saltpetre, which is poison to poultry. While they are very young, keep them wrapped up constantly, only taking them out for a short time to feed. As they grow bigger, gradually increase their liberty, so that by the time they are ten or twelve days old, they may run about all day in a sunny verandah if they are under a box with a wire top.

"When older, they can be banished to the *murghi khana*. The Only disadvantage to this way of rearing chickens is that they become so tame that there is no keeping them out of the house, and it is painful to order such tame creatures to be killed for the table. Poultry reared in this way are generally quite free from insects and contagious diseases. This is probably due to the fact that insects and germs of disease hang about the feathers of old hens.

Chapter 9

Incubators

Those fortunate enough to possess an incubator soon learn to use it successfully; and, consequently, anyone in earnest about the management of poultry should go to the expense of getting one. No incubator taking less than 50 eggs is worth having, for this number will hatch with the same time and trouble as any smaller number. These clever contrivances do away with all that is disagreeable and tiresome in setting hens, who may be wayward, only one out of three, at the most, bringing out a successful brood. Those who are expert in managing incubators can easily contrive to make them up, with the exception of the capsule and thermometer which should be procured from England. In the books of instruction which accompany incubators, there is much nonsense regarding the management of the chickens, who, if healthy, only require a little care and common sense treatment. Christy's incubator appears to give the best results in the plains, and Pearson's in the hills, as it keeps a more equable temperature, and does not require so much attention.

Chapter 10
(The Chapter of Chapters)

How to Kill and Dress a Fowl Properly So that it may be Pleasant to the Taste, and Wholesome

In the morning take a well-fed fowl before it has had its breakfast, tie its legs together; hold its head so as to cover the whole of it, leaving the neck free; swing the body round, making a pivot of the point where head and neck meet. As the body swings round, the neck is broken, and the fowl is killed without suffering pain; it flutters slightly for a few seconds, but that is all. The sweeper readily learns this mode of killing. The table servants, on religious grounds, will *not* do it. No one who has been in the habit of eating properly killed poultry will touch a fowl slaughtered by the cook in the orthodox fashion. In England all poultry is killed by men whose business is to do it, and is never left for the cook to mutilate. In this country it is incumbent on all house-keepers to insist on their poultry being properly killed, in order to avoid both unnecessary cruelty and the preparation of unwholesome and unpalatable food.

Before the bird is cold, pluck it; this is easily done by pulling the feathers upwards a good plan is to hang it up by the legs while removing the feathers; they have it drawn, and the inside wiped out with a clean cloth. Should the weather be cold hang up the bird for a day or two before plucking and drawing; this helps to make it tender. If possible, all birds should be hung. Should the bird be a large one, such as a goose or a turkey, have it shot through the head instead of having the neck broken. When the bird has been plucked, drawn and wiped out, stuff the whole body with stuffing made as follows:

> I part Bread, I part Suet flavoured with a little minced Ham or Bacon; Thyme, Mint or Sage, Pepper and Salt; the whole mixed up with an egg and some milk.

The body of the fowl having been fully stuffed should be carefully sewn up at both ends, So as to prevent any of the stuffing oozing out. In cooking, the stuffing swells up, and its goodness rises through the bird, and makes it plump and juicy.

If the bird is to be boiled, place it in a *degchee* with a little water-sufficient to cover an inch of the bird–cake down closely with dough round the edges of the cover, and cook *very slowly*. Should soup be in course of preparation, boil the fowl on the top of the soup meat, letting it boil only in the steam rising from the soup.

If the fowl is to be roasted, after stuffing and careful sewing up, smear it over with a well-whipped egg, dredge with flour and a little salt, and roast in a stove or in a *degchee*.

A fowl or duck, killed and cooked as above, will be delicious, both hot and cold.

Pullets, after laying their first eggs, are excellent eating if killed as soon as they become brood; a layer of fat will be found under the skin, as nature provides a hen with extra fat and flesh at this time, to enable her to sit out 21 days, at the end of which she is thin and dry.

Feather Pillows

All the small feathers should be carefully collected, and dipped into a pail of water in which pounded alum has been dissolved. After a day or two, strain off the water, and dry the feathers in a muslin bag. Feathers so treated make good pillows: feather pillows are difficult to get in India.

Table of Age for Killing Poultry

Country hens should be killed as soon as they have laid their first batch of eggs; country cocks as soon as they crow; English hens in their fourth year; English cocks when 5 years old.

English fowls are always tender, but the cocks are useful in the yard till they are five.

Chapter 11
Diseases of Fowls and Chickens

Cramp

Rub the limbs with equal parts of kerosene and mustard-oil.

Give, morning and evening, the following mixture:

1 Tea-spoonful pounded red pepper.

1 Tea-spoonful ginger.

1 lb *atta*.

Mix the above dry, and when required, mix about a teaspoonful with mustard-oil, and give it in pills. Chickens should have less.

Weak Legs

This is a sign of outgrowing strength, and generally attacks young cocks. They require animal food. Feed with raw meat and a little bone-dust mixed with *atta*. Should the bird be a valuable one, give it either Parish's Chemical food or Fellow's Syrup of the Phosphites–a few drops daily; but, as a rule, meat diet soon strengthens the legs.

Scaly Legs

This is a very common complaint among English fowls, particularly those with feathered legs, and it is apt to run on unobserved until the poor bird gets large scales all over its legs and toes, and eventually becomes lame and useless. With a little perseverance it is easily cured. It is caused by a small parasite which lives in the skin, and the scales grow from the constant irritation, and make still better shelter for the insects, so the longer the complaint is neglected, the worse it gets. The legs should be dipped, morning and evening, in kerosene oil. A less troublesome method is to paint them with carbolic and oil with a small amount of castor in it. The legs will at first, look raw and sore, but soon get well. Another cure is turpentine. The complaint should be treated, as soon as observed, for, being due to a parasite, it quickly spreads among poultry.

Bumble Foot

A complaint almost entirely confined to old poultry; it makes the birds perfectly lame and helpless. Paint with iodine daily; if there is no improvement, rub with caustic twice a week. If the tumor continues to increase and grow harder, make a deep incision across it, and squeeze out the contents: let out as much blood as you can. The complaint is hereditary, and very persistent and troublesome.

Liver

There are two forms of liver complaint–decayed liver and enlarged liver. The former is caused by sleeping in hot close pens at night, and more often by damp and wet; cocks are more subject than hens to this complaint, which is an insidious one, at first presenting no symptoms of deranged health beyond white droppings. At this stage either kill the bird for the table, or treat as follows: Chiretta leaves pounded up and mixed with *atta*–one; pill, as big as a nut, morning and evening–keep the bird in a dry place. It has probably had a chill, which will, in nine cases out of ten, kill it, if not very carefully treated. If the course of the disease is not checked, the face and mouth will become pale in colour, the eye looks savage and keen-a look you cannot mistake–the bird gets daily thinner, the legs grow weak and the diarrhoea worse, till the droppings are of a green colour, after which the bird will die. A post mortem examination will show that the liver has almost disappeared. This complaint is common to birds which have been improperly fed when chickens; feeding with rice will also produce it. Hens often get it by continuing brood, and sitting on for weeks in a hot, close place.

Enlarged liver is a common complaint in the hills, where birds, apparently in perfect health, droop for a few hours and then die. On opening the bird, the organs will be found quite healthy, but the liver is big enough for a goose, and has, at length, pressed on the heart and stopped its action. A fowl with an enlarged liver is quite fit for food, so, if you find that the birds are affected with this complaint, kill them off for the table. It is caused by sleeping in houses which are hot and close, and by standing on damp ground. The best preventive is a high perch. Only English fowls, as a rule, contract this disease.

Death From No Apparent Cause

Sometimes fowls die off in numbers without any apparent cause. This is due to damp; they have probably been allowed to get wet, or to run about on damp ground. If, during the rains, they are kept closely confined in a dry house, they will never die off in this disheartening way. Damp and wet are fatal to poultry.

Loss of Feather

This should not occur when the birds are not moulting. If it does, they are out of health, and require green food and a tonic. Those affected should be removed from the others, who will pick at the bare places and make them worse. A bath of warm water, in which is carbonate of ammonia, will, if the weather is dry, prove of service. A bird with bad plumage is unfit for the table, as its flesh is unhealthy.

Moulting

With well-bred fowls this is always an anxious time. Generous diet should be provided and hot mashes. Old birds often have difficulty in casting off the old feathers, and, if not looked after, sometimes die from the inflammation caused by the congested secretion. If the birds look wretched while moulting, pluck out every old quill you can find; they are brown and dry, and very difficult to pluck out; after this has been done, the bird will be relieved at once, and, in a few days, a new strong crop will fill up the holes from which the dead feathers were pulled out.

Should a hen shed its old quills, and no new ones be observed, give it a tonic, and also a hot bath. with a little carbolic in it.

The health of poultry should not be affected during moulting; the cocks leave off crowing, and the hens laying and in a little time it is seen that they have lovely glossy plumage in place of what had become tumbled and soiled.

Drooping Wings

This complaint is a sign of debility, and is common in chickens in the rains, particularly if they get wet. All the heavy feathers should be cut off, and the bird kept dry. Give it a teaspoonful of whiskey morning and evening. Feed on mash, and let it have an iron tonic. It will soon pick up again.

Soft Eggs

These are eggs without a proper shell; apparently they have only the lower skin, The hen requires lime; give it mixed with a little *atta*. Throw the mixture among the hens, and put a little of it into their drinking water.

Egg Bound

The hen goes about, at laying time, with drooping wings and in great distress. She should be taken up very carefully and examined, If the egg is still soft, and far back, salad oil should be inserted with a feather or small syringe, and the hen held over warm water, so that the steam may expand the part, and allow the egg to pass, If the egg is in the passage, it is a dangerous state of affairs, and if not at once attended to, the result will be fatal. It is caused by hens being bred up to lay larger eggs than nature intended; also by constipation, and by a double-yolked egg. The egg, while being laid, is soft, as it is before it is laid, and hardens as soon as it is exposed to the air; if, therefore, there is an unnatural delay in expelling it, it hardens in the passage. Should the hen be found struggling in the nest, and apparently dying, the part should be at once well oiled, and pressure put all round to help out the egg, but if unsuccessfully, a small hole should be made in the egg, and the contents allowed to run out, the egg may then be taken out piecemeal with the little finger. Native women, whose hands are small and clever, perform this operation well. Care must be taken that no particles of egg-shell are left, or the hen will die of internal inflammation. If the operation is successfully performed, the bird, in half an hour's time, will be running about as lively as ever. The next day a little oil should be applied, to prevent a recurrence.

Broken Eggs

This is a more serious occurrence than the former, as very often it is not discovered. The hen seems distressed and ill; she is rough, and her eye has a bright savage look. She will look at food, but will not lower her head to pick it. She should be examined, and if a discharge is observed resembling the yolk of an egg, the cause is clear. Syringe at once with warm water until everything is brought away; then syringe with castor-oil. Repeat for a day or two to make certain that nothing is left, or inflammation will be set up. It. is caused by rough, clumsy cocks attacking the hen just as she is going to lay, or by being held roughly by the sweeper, or caught by being pounced upon, and struck violently against the ground.

Hard Crop

Sometimes, in even quite healthy poultry, a bird will suddenly get a swelled, hard crop. The gram in it begins to germinate the bird is much distressed, and cannot touch food. A slight examination will show what is wrong, and an incision about an inch long must be made in the middle of the crop, avoiding a blue vein–the contents of the crop must be removed, and an examination made with a well-greased finger, to see if the lower valve is choked up, as this will sometimes produce crop-binding. When you are satisfied that the passage is clear and the crop quite empty and clean, give the cut three stitches, and keep the bird shut up. Feed it, for a few days, on soft meal, but very sparingly; the wound will heal quickly, and the bird recover.

Soft Crop

This is a troublesome complaint, not easily cured. The crop is swelled, soft and pulpy. The bird drinks freely, and ejects much that it has drunk. The crop becomes offensive, and the bird appears sickly. Mrs. Bellairs, in her little book, recommends charcoal pills. Peppermint and castor oil are useful, as the crop is full of wind. I also give a little brandy or whiskey. This complaint often arises from drinking dirty water, especially water in which ducks have dabbled.

Dysentery

Give soft food, and, three times a day, pills made as follows:

Opium	3 grs.
Catechu	3 grs.
Made into 12 pills.	One pill for a dose.

Poisoning

If poisoning is suspected, give castor-oil three times a day in large doses. When the bird is recovering, add laudanum to the castor-oil. Continue as long as any signs remain in the droppings, such as slime, blood, &c.

Scouring and Straining

Six drops of chlorodyne. Give in a tea-spoonful of salad-oil and also hot food.

Scrufy Face and Comb

A contagious complaint due to parasites. The parts should be painted with equal parts of sulphuric acid and glycerine.

White Comb

A comb which is white, or of any colour except bright scarlet, denotes bad health. A general attention to health is therefore indicated. Either green meat, animal food, iron or tonic may be necessary.

Black Comb

This is a serious disease, and kills very quickly. I believe that in Holland it is called *Black Rot*. The comb takes a blackish purplish colour, and the legs are affected. It can only be cured if observed in the earlier stages, when the bird appears well in other respects. Feed well, and give a tonic, with an occasional dose of calomel and jalap.

A fowl suffering from even a slight attack of this complaint is unfit for the table, as its flesh is, or is inclined to be diseased.

Colds and Coughs

Colds are easily discovered by looking at the nostrils; if they are moist, and thin discharge flows from them, and there is a watery look round the eyes, the bird has a cold, generally contracted by standing on wet ground, or perhaps due to gross condition. Give aniseed, called by natives *Somph*, which is in every bazaar, and very cheap. A lump of camphor, about as big as a pea, put down the throat every morning, is also efficacious. In fact, camphor is a specific in many complaints, and is invaluable when epidemics are about. A lump of camphor given to each fowl will often prevent infection.

See, later on, for Roup, Chicken-pox &c.

Fever

Should a bad chill take the form of fever, which it frequently does in the plains during the rains, the symptoms will be as above, but in addition the face will be red and hot, and the breath feverish. Give one grain of quinine with two grains of camphor morning and evening, and the bird will soon recover if the fever is of an ordinary type, but should it be sickening for roup or chicken-pox, it will be much worse on the third day; it is best, therefore, to isolate a bird with a cold until it has recovered.

Inflammation

Sometimes inflammation is set up in the egg-producing organs of a bird which is too fat. If so, kill her for the table, as her eggs will not be fertile, and she will not lay many.

Should inflammation of an inner part occur after roup, the bird will become thin and light, and is quite unfit for food. She should be dosed with laudanum varied with a little castor-oil.

Castor-oil

Some poulterers strongly object to this medicine, preferring Epsom salts or jalap. Lewis Wright says: "More poultry have been killed by it than by any other medicine." I do not like to combat his opinion, but must admit that I have found it very efficacious in this country. Perhaps castor-oil, as prepared in England, may be harmful to poultry, but not the raw oil obtained in Indian bazaars, which is thick, dark coloured, and pure, and freshly made from the seeds. It is more powerful than English oil, and consequently goes further, and it does not cause subsequent constipation.

Epsom Salts and jalap are useful in many complaints, but castor-oil should be given if the bowels are in an inflamed state from poison or other causes. If it is too strong, mix with it an equal quantity of salad oil.

Gapes

A complaint more common in England than in India. It is caused by a reddish worm formed in the chicken's throat, and it gapes for breath as it suffers from suffocation. Dislodge the worms with a feather stripped nearly clean, and then paint with a weak solution of caustic.

Insects

Five different kinds plague poultry: "Hundreds-and-thousands," wood-bugs, house-bugs; ticks, and lice.

Poultry in the plains sometimes suffer from the attacks of all five at the same time, especially the poor *murghies* belonging to natives.

"Hundreds and Thousands" are very small, and of a yellowish colour. They are almost invisible to the naked eye, but under a magnifying glass they look like fat-bodied-spiders. They plague the poultry in February and March. They so trouble the hens that they forsake the nest, and if the insects swarm on the newly-hatched chickens, the latter will die. These pests are easily discovered, as they run on to human beings, and cause a tickling sensation on the skin. If they are too small to be seen readily, their presence may be suspected if the fowls keep shaking themselves, and rustling their feathers.

To destroy them spread dry leaves or woodshavings over the floor of the fowl-house, and set fire to them. the flames should be allowed to run two feet up the walls, above which the insects do not go, remaining chiefly on the floors, where they deposit myriads of eggs. All straw and baskets should be burnt.

Brood hens should not be allowed straw in their nests, as it harbours these insects. Dry saw-dust is best, and if the floor is covered with a thick layer of dry earth, the hen's will be pretty clear of them. Sprinkle the sitting hens with Keating's Insect Powder, which destroys these vermin.

Wood-bugs, Otherwise Wood-lice

These disgusting insects live in wood and in trees. They infest nearly every fowl-house in the plains, and increase in thousands.

By day they live in cracks in plaster under bricks, in all wood-work and bamboos, in fact in every place in which shelter is to be obtained. At night they run up the perches and the legs of the poultry, causing, by their bites, rawness and redness. The birds will become weak, pale and sickly, and the hens will leave off laying. These insects are oblong in shape, brown in colour, and dry- looking, resembling wood. They are of all sizes up to half an inch in length. If touched, their legs curl up, and look like little chips.

Clean and re-plaster the fowl-house. The best plan, if practicable, is to remove the poultry for a month, when the insects will die of starvation. Painted roosts are also a protection. Wood-bugs destroy ducks.

House-bugs

These bugs infest fowl-houses in the hills, especially if they adjoin the servants' quarters. House-bugs are easily discovered by examining the cracks in the perches and laying boxes. The latter should have boiling water, with phenyle in it, poured over them; or they should be burnt, and new boxes provided.

Ticks

These are to be found, in patches of all sizes, on the skin, under the feathers; they look like brown beads, and the skin all round is red and inflamed. They spread quickly, and soon reduce the bird. They adhere so firmly to the skin, that they cannot be picked off. They are difficult to get rid of, as-each patch or colony of them must be painted with a mixture of phenyle and salad oil. If the smallest patch escapes the brush, the bird will soon be covered again. These creatures go on increasing until they kill their victim. Painting with kerosene-oil is also a cure, as it kills the parasite.

Lice

These live in the feathers, and are easily discerned. They only breed in very dirty and badly- kept poultry. Sometimes a bird, who has been ill for a long time, and is sickly, may be swarming with them, while its healthy companions are free.

They are destroyed by rubbing in white precipitate dissolved in vaseline.

Roup

Roup is terribly infectious, and is, consequently, one of the worst diseases to which a poultry-yard is subject. The infection is carried not only by chickens and fowls, but also by buckets and pans, and even by woollen clothes. I am not a doctor, so my statement on medical matters may be open to question, but roup looks to me like a kind of diphtheria, and I should like to see it subjected to the anti-toxin treatment.

Symptoms

Observe the bird's throat, and if the part under the bill is moving up and down, as if there was a difficulty in breathing, open the bill and look in. If the wind-pipe over the tongue is distended and white, the fowl most certainly has roup. In a healthy throat the wind-pipe is so far back, and is so small, that it is not easily seen; but when attacked by roup, it stands up like a bit of white India-rubber tubing, showing the breathing-valve distinctly; the whole the throat and the lining of the bill will be of a white dusky hue, and the mouth will have white glutinous matter hanging about it; if the disease is in an advanced stage, the respiration will be slow, and a whistling sound will be produced as the breath passes through the windpipe; the fowl will suddenly sneeze and jerk its head, as if trying to get something out of its throat, and then it will again pull a long breath, making a whistling sound as it fills its lungs. The most reliable symptom is the odour of the breath. From the first this is most offensive, and an experienced person will distinguish it at once.

This serious disease sometimes appears suddenly, and at other times the progress is slow. Some birds suffer from a chronic type, and will live on for months with a white flaky throat and perpetual sneezing, as the disease has travelled inwards, and has only slightly attacked the throat. These fowls will spread the disease wherever they go, although they themselves do not die for a long time; they will move about, lay soft eggs, and remain sickly. If opened after death, the white fungus-like growth will be found to have spread all through the crop, bowels and egg-producing membranes, and even to have extended to the skin. Another symptom of the disease is thirst. At night, too, when roosting, the whistling sound is heard.

This complaint is generally supposed to be incurable, but, as the treatment I adopt has been very successful. I give it here in the hope that others may find it the same–Give a lump of camphor, as big as a pea, three times a day. Feed on a hot mash made of *atta* and hot fat. Twice a day administer a tea-spoonful of hot whiskey and water. Remember that the camphor is indispensable. If the throat and mouth are covered with the white fungus, burn it out with a stick of caustic, and give the bird a little bit of caustic to swallow, which will clear the growth from the crop and bowels.

On the first suspicion of roup in your yard, give every fowl a large lump of camphor. It is perfectly harmless. You need be afraid of giving too much--and it certainly prevents the disease from spreading. I have only failed in my cure when the complaint has not been observed till the fungus has spread inside, and the bird has become very weak. Many fowls die of weakness after the disease has been cured, so feed generously on melted fat, *ghee* and hoof-oil (the skimmings from boil trotters). No solid food should be allowed, as the digestion is too weak for it; pounded raw meat and whiskey may be given.

In some cases, after the cure of roup, a swelling will rise on the face, which becomes white and distended. This should be opened, and out of it will come a large lump of white spongy substance, resembling a button mushroom. If this is removed entirely, the swelling will subside; otherwise, it will grow to such a size that the poor bird will become blind, and ultimately die from the mouth becoming choked up.

A fowl suffering from roup should be isolated, and kept far from all other poultry, as the discharge from the mouth and nostrils is very infectious. When the bird has been cured, the pen in which it has been confined should be burnt, otherwise the complaint may be reproduced in damp weather.

Roup is entirely and wholly caused by infection, and will alike attack poorly and well-fed poultry. Neither good feeding nor newly-built houses, nor dry roosts, nor any other comfort will prevent its breaking out, if the birds are exposed to infection, which may be brought by a stray fowl, or even by a seller of poultry seating himself in the vicinity of the yard. An old basket, or a sweeper's broom, may convey the germs. It is worst in damp weather, and for this reason many have supposed that damp is a cause, but it is only an exciting cause. The germs must be present first, and damp helps them to activity. If roup has been in the yard one year, it is very likely to re-appear next year as soon as wet weather comes.

When roup attacks a country fowl, it is hardly worth while to attempt a cure. Kill the bird and get rid of it, and dose the remainder with camphor–but it is different in the case of a valuable English bird.

The disease is not conveyed to human beings, but the discharge will cause bad sores on the hand, if care is not taken to prevent its getting into a cut. Bazaar fowls and chicken are often in an advanced stage of roup when bought cheap by *khansamahs*, who carry them home for their innocent masters, who do not suffer any ill effects from such food although, properly speaking, they should get roup!

Chicken-pox

It is so called because it attacks chickens more frequently than full-grown fowls. It is an epidemic, and very infectious. It is not so common in the hills as in the plains, where spring and autumn are the usual seasons for it. It clears off whole brood, of young chickens; many who recover are lame and blind for life. Very often the inflammation settles in the ball of one eye, and if the eye is taken out, the bird is left healthy.

The first symptoms are observable in the face, which becomes red and inflamed. The head feels hot to the hand. The eyes are watery, and the nostrils moist, as if the bird had a bad cold. The tail droops, the back is rounded, and the head draw-back into the breast. The next day the eyes will remain closed, and the face will be more swollen. On the third day ulcers will be observed, which will go through all the regular stages of chicken-pox. The head becomes a mass of sores and scabs, the bird becomes blind, and soon the bill will be forced out of shape.

The best remedy is to bathe the face with Condy's fluid, and give a little of the same to drink. Natives recommend mustard oil, pepper-corns &c., but they increase the temperature, and cause the fever to run higher. Cooling food should be given–bread and milk, bread and butter, *suttoo* made into pills and given by hand. Powdered charcoal mixed with cocoanut-oil should be sprinkled over the sores if they smell bad, or paint with phenyle. Turkey chicks are especially subject to this complaint, and it spreads very quickly among them. If one bird in a poultry-yard gets it, the probability is that all will be attacked, excepting those who may have had it before.

Swelled Face

It resembles chicken-pox in the early stages, but if, after the third day, no pocks appear, it is due to a feverish cold. Bathe the head with warm spirit and water; give a few drops of camphorated spirits and quinine three times a day. The complaint is rapid in its development, and if a hen is attacked badly, it may run to a fatal termination in a couple of days. It is not infectious, but many fowls may suffer from it at the same time, as they all are exposed to the same climatic influence. Natives bathe the head with strong salt and water and syringe the throat with the same. Keep the bird in a warm place and feed on mustard-oil mixed with *atta* and a few pepper-corns. On recovery give quinine.

Cramp and Diarrhoea, in the Hills

Attack fowls brought from the plains to hills. They are due to change of climate, and many birds die. On arrival they should, for the first week, be fed on warm food–*atta* mixed with hot water and mustard-oil. This should be given twice a day, the first thing in the morning and the last at night, to keep off chills till the bird is acclimatized. Generous feeding with raw meat is also recommended.

Index

Cow

Poultry